Soil salinity assessment

Methods and interpretation of electrical conductivity measurements

FAO IRRIGATION AND DRAINAGE PAPER

57

by

J.D. Rhoades

US Salinity Laboratory

United States Department of Agriculture

Riverside, California, USA

F. Chanduvi

Water Resources, Development and Management Service

FAO Land and Water Development Division

S. Lesch

US Salinity Laboratory

United States Department of Agriculture

Riverside, California, USA

Food and Agriculture Organization of the United Nations

Rome, 1999

The designations employed and the presentation of material in this publication do not imply the expression of any opinion whatsoever on the part of the Food and Agriculture Organization of the United Nations concerning the legal status of any country, territory, city or area or of its authorities, or concerning the delimitation of its frontiers or boundaries.

M-51
ISBN 92-5-104281-0

Foreword

The technology described in this report for measuring soil salinity has been extensively and successfully field-tested. It is concluded to be sound, reliable, accurate and applicable to a wide variety of useful applications. It is based on proven theory of soil electrical conductivity. The required equipment is commercially available. The advocated instrumental methodology is practical, cost effective and well developed for essentially all general applications. It is cheaper, faster and more informative than traditional methods of salinity measurement based on soil sampling and laboratory analyses. Software is available to facilitate its use for mapping and monitoring uses, as is equipment to mobilize and automate the measurements for use in detailed field-scale assessments. Its usefulness has been demonstrated: 1) for diagnosing soil salinity, 2) for inventorying soil salinity, 3) for monitoring soil salinity, 4) for evaluating the adequacy and appropriateness of irrigation and drainage systems and management practices, 5) for determining the areal sources of excessive leaching, drainage and salt-loading in crop lands, 6) for establishing the spatial soil information needed to develop prescription farming plans to manage fields with spatially-variable salinity conditions, and 7) for scheduling and controlling irrigations under saline conditions. It offers the potential to identify the inherent causes of salinization in fields, especially when integrated with GIS technology, and to identify mitigation needs, especially when integrated with field-scale deterministic, solute-transport models. The salinity assessment approach advocated in this report offers a more suitable basis for evaluating, managing and controlling soil salinity than do the leaching requirement and salt balance concepts/measurements as traditionally applied. National programs need to be implemented to mitigate the substantial problems of secondary salinization that threaten the sustainability of irrigation in many places in the world. Holistic; meaningful salinity assessment approaches needed in this regard are illustrated in this report. The presented salinity assessment technology offers substantial practical potential to inventory, monitor, manage and control soil and water salinity, as will be needed to sustain irrigated agriculture and to meet the worlds food needs in the coming decades.

Acknowledgements

The authors wish to acknowledge the valuable advice provided by the staff of the Land and Water Development Division, who revised and contributed to the general outline of the publication. The technical reviews of the first draft by Messrs N.K. Tyagi (Central Soil Salinity Research Institute, Karnal, India) and J.W. van Alphen (International Agricultural Centre, The Netherlands) are gratefully appreciated.

Most of the research findings and recommendations in this publication come from the United States Salinity Laboratory, Riverside, California, and the authors acknowledge the Laboratory staff for their valuable contribution.

The authors with to express their gratitude to Dr Arumugan Kandiah, Programme Manage of the International Programme for Technology and Research in Irrigation and Drainage (IPTRID) of FAO for his technical and overall support towards the fulfilment of this publication.

Thanks are also due to Ms C.D. Smith Redfern for her valuable assistance in the preparation of the final camera-ready text and figures and tables. She also contributed to the improved presentation of the document.

Contents

List of figures

Page

Symbols

ε	a random error component
β	a regression-fitted parameter estimate
α	depth-weighted mean anion exclusion volume as a proportion of the volumetric soil water content above depth z in the soil
λ	the ratio of the solute concentration at depth z to the mean concentration; s_z/s_m
ρ_b	bulk density of the soil
°C	temperature in degrees Celsius
Θ_e	volumetric content of soil water of a saturated paste
Θ_{fc}	volumetric content of soil water at "field-capacity"
Θ_m	mean volumetric content of soil water averaged over depth z
ρ_s	density of the soil particles
Θ_s	volumetric content of the solid phase of soil which does not contain an indurated layer
Θ_{sc}	volumetric content of solid phase of soil (which contains an indurated layer) in a continuous electrical solid-pathway
Θ_{ss}	volumetric content of solid phase of soil (which is contains an indurated layer) in a series-coupled electrical solid-pathway
ΔS_{sw}	the change in soil solution salinity within the rootzone depth
Θ_w	volumetric content of total soil water
Θ_{wc}	volumetric content of soil water in the continuous-liquid pathway; essentially that in the larger, continuous pores ("mobile water")
Θ_{ws}	volumetric content of soil water in the series-coupled pathway; essentially that in the small pores & films ("immobile water")
b_z	proportion of applied water moving as by-pass flow past depth z in soil
c	concentration of tracer solute in the irrigation water
C_{dw}	concentration of salt in drainage water flowing from the rootzone
C_{gw}	concentration of salt in shallow groundwater
C_{iw}	concentration of salt in irrigation water
C_{tw}	concentration of surface runoff water (tailwater)
EC_{25}	electrical conductivity referenced to a temperature of 25 °C
EC_{4p}	EC_a measured using four-electrode array
EC_a	electrical conductivity of bulk soil
EC_a^*	depth-weighted value of EC_a
EC_e	electrical conductivity of the extract of a saturated soil-paste
EC_s	electrical conductivity of surface conductance of soils without indurated layer
EC_{sc}	electrical conductivity of indurated solid phase
EC_{ss}	electrical conductivity of surface conductance of soils with indurated solid phase
EC_t	electrical conductivity at the temperature of the sample

EC_{wc}	electrical conductivity of Θ_{wc}
EC_{ws}	electrical conductivity of Θ_{ws}
EC_x	EC_a within a depth-interval, as estimated from a sequence of surface-array four-electrode readings
EM	electromagnetic induction
EM_H	EM measurement made with the axis of the magnet-coil in the horizontal configuration
EM_V	EM measurement made with the axis of the magnet-coil in the vertical configuration
f_z	proportion of flow that occurs at the concentration of soil water
I	rate of irrigation averaged over period of measurement or calculation
L	leaching flux at depth z averaged over period of measurement or calculation
L_f	leaching fraction; V_{dw}/V_{inf}
log	the logarithm of a value to the base ten
S_c	amount of salt removed from soil solution by crop uptake
S_f	amount of salt added to the soil solution by the dissolution of fertilizers and amendments
S_m	amount of salt brought into soil solution by mineral weathering
S_m	mean concentration of conservative solute averaged over depth z
$S_{m(0)}$	mean concentration of conservative solute averaged over depth z at time t_0
S_p	amount of salt removed from soil solution by the precipitation of salt-minerals
SP	saturation percentage; the gravimetric water content of a saturated soil-paste
s_z	mean concentration of tracer solute at depth z averaged over period of measurement at the reference water content Θ_m
T	[temperature in degrees Celsius - 25] /10
t	time period of measurement or calculation
t_o	initial time of measurement, or reference time for beginning of calculation
V_{cu}	volume of water consumed by crop in evapotranspiration
V_{dw}	volume of drainage water flowing from the soil rootzone
V_{gw}	volume of groundwater which flows into the rootzone depth of soil
V_{inf}	volume of infiltrated water (V_{iw} - V_{tw})
V_{iw}	volume of applied irrigation water
V_l	volume of net leaching (V_{dw} - V_{gw})
V_{tw}	volume of surface runoff water (tailwater)
W	a trend surface matrix based on the spatial coordinates of the measurement sites
X	a matrix of log transformed and de-correlated sensor readings
Y	the vector of log transformed soil salinity values
z	depth in soil profile

Chapter 1

Introduction

The world's demand for food is increasing at such a rate that the ability to meet anticipated needs in the next several decades is becoming questionable. Irrigated agriculture presently accounts for about one-third of the world's production of food and fibre; it is anticipated that it will need to produce nearly 50 percent by the year 2040 (FAO, 1988). This will likely be difficult, because extensive areas of irrigated land have been and are increasingly becoming degraded by salinization and waterlogging resulting from over-irrigation and other forms of poor agricultural management (Ghassemi, *et al.*, 1995). Available data suggest that the present rate of such degradation has surpassed the present rate of expansion in irrigation (Seckler, 1996). In some places, the very sustainability of irrigated agriculture is threatened by this degradation (Rhoades, 1997a; Rhoades, 1998). At the same time, irrigated agriculture is also depleting and polluting water supplies in many places. Increased irrigation efficiency is being sought to conserve water, to reduce drainage, waterlogging and secondary salinization, and to mitigate some of the water pollution associated with irrigated agriculture. Restrictions are increasingly being placed on the discharge of saline drainage water from irrigation projects. Concomitantly, the reuse of saline drainage water for irrigation is being increased. With less leaching and drainage discharge and greater use of saline water for irrigation, soil salinity may increase in some areas. Thus, a practical methodology is needed for the timely assessment of soil salinity in irrigated fields, for determining its causes and for evaluating the appropriateness of related management practices.

Ideally, it would be desirable to know the concentrations of the individual solutes in the soil water over the entire range of field water contents and to obtain this information immediately in the field. Practical methods are not available at present to permit such determinations, although determinations of total solute concentration (i.e., salinity) can be made in situ using electrical or electromagnetic signals from appropriate sensors. Such immediate determinations are so valuable for salinity diagnosis, inventorying, monitoring and irrigation management needs that, in many cases, they supplant the need for soil sampling and laboratory analyses. However, if knowledge of a particular solute(s) concentration is needed (such as when soil sodicity or the toxicity of a specific ion are to be assessed) then either a sample of soil, or of the soil water, is required to be analysed. Of course, the latter methods require much more time, expense and effort than the instrumental field methods. In this case, a combination of the various instrumental and laboratory methods should be used to minimize the need for sample collection and chemical analyses, especially when monitoring solute changes with time and characterizing the salinity conditions of extensive areas.

Customarily, soil salinity has been defined and assessed in terms of laboratory-measurements of the electrical conductivity of the extract of a saturated soil-paste sample (EC_e;

this as well as all other symbols used in this report are summarized in the list of symbols). This is because electrical conductivity is an easily measured and practical index of the total concentration of ionized solutes in an aqueous sample. The saturation percentage (SP) is the lowest water/soil ratio suitable for the practical laboratory extraction of readily dissolvable salts in soils (US Salinity Laboratory, 1954). As the water/soil ratio approaches that of a field soil, the concentration and composition of the extract approaches that of soil water. Soil salinity can also be determined from the measurement of the electrical conductivity of a soil-water sample (EC_w). This latter measurement can be made either in the laboratory on a collected sample or directly in the field using in situ, imbibition-type salinity sensors. Alternatively, salinity can be indirectly determined from measurement of the electrical conductivity of a saturated soil-paste (EC_p) or from the electrical conductivity of the bulk soil (EC_a). EC_p can be measured either in the laboratory or in the field using simple and inexpensive equipment. EC_a can be measured in the field either using electrical probes (electrodes) placed in contact with the soil or remotely using electromagnetic induction devices. The latter two sensors are more expensive than those used to measure the EC of water samples, of soil-extracts or soil-pastes. However, their use is very cost effective when one considers the amount of spatial information that can be acquired with them (the relative costs of the different methods for assessing soil salinity are discussed later; the basis for this economic evaluation is presented in Chapter 5). From measurements of EC_p and EC_a, soil salinity can be deduced in terms of either EC_e or EC_w. The appropriate sensor and method to use depends upon the purpose of the salinity determination, the size of the area being evaluated, the number and frequency of measurements needed, the accuracy required and the available equipment/human resources.

Traditionally, the leaching requirement (L_r) and salt-balance-index (SBI) concepts have been used to judge the appropriateness of irrigation and drainage systems and practices, with respect to the avoidance of salinity and waterlogging problems; these concepts have also been used to estimate the extra water requirements associated with saline irrigated lands (US Salinity Laboratory Staff, 1954). However, these approaches are either inadequate or impractical for these purposes. The leaching requirement (L_r), refers to the amount of leaching required to prevent excessive loss in crop yield caused by salinity build-up within the root zone from **the salts applied in the irrigation water**. Its calculation is based on the assumption of steady-state and of uniform conditions of irrigation, infiltration, leaching and evapotranspiration; none of which are achieved in most field situations which typically are dynamic and variable, both spatially and temporally in the above mentioned attributes. Furthermore, salt build-up in the root zone resulting from the presence of shallow water tables is ignored in the traditional L_r calculation. Additionally, no practical way has existed to directly measure the degree of leaching actually being achieved in a given field, much less in the various parts of it, as is required in order to determine its appropriateness. However, a potential means has been developed to estimate the extent and adequacy of leaching based on measurements of the levels and distributions of salinity within irrigated root zones, as will be described later.

The salt-balance index (SBI), which is the net difference between the amount of salt added to an irrigation project and that removed in its drainage effluent, is another "concept" that traditionally has been used to evaluate the appropriateness of irrigation, leaching and drainage practices. This approach is also inadequate for these purposes because it provides no information about the average level of soil salinity in the project, nor about the actual level of soil salinity existing within any specific field of the project. The approach also fails because it does not even provide a realistic measure of trends in salinity **within the root zone,** because salt derived from below the soil profile and of geologic origin is typically contained in the drainage water collected by the subsurface drainage system (Kaddah and Rhoades, 1976). Additionally, the transit times involved in the drainage returns are so long (often more than 25 years) that the index values are not reflective of current conditions/trends (Jury, 1975a, 1975b). From project-wide SBI values, one can not deduce the extent of leaching being achieved in any field, nor the uniformity and efficiency of irrigation and leaching, nor the extent of waterlogging- and of salinity-induced losses in crop yield; traditional SBI measurements are impractical to make on the basis of individual fields and of root zone environments.

In the author's opinion, the effective control of soil salinity and waterlogging, and also of salinity in drainage-receiving waters, requires the following: (i) knowledge of the magnitude, extent and distribution of root zone soil salinity in representative fields of the irrigation project (**a suitable inventory of conditions**); (ii) knowledge of the changes and trends of soil salinity over time and the ability to determine the impact of management changes upon these conditions (**a suitable monitoring programme**); (iii) ways to identify the existence of salinity problems and their causes, both natural and management-induced **(a suitable means of detecting and diagnosing problems and identifying their causes**); (iv) a means to evaluate the appropriateness of on-going irrigation and drainage systems and practices with respect to controlling soil salinity, conserving water and protecting water quality from excessive salinization (**a suitable means of evaluating management practices**), (v) an ability to determine the areas where excessive deep percolation is occurring, i.e., to identify the diffuse sources of over-irrigation and salt loading (**a suitable means of determining areal sources of pollution**), (vi) knowledge of the spatial variability in soil salinity needed to develop site-specific management "tailored" to deal with such variability and to avoid excessive and wasteful inputs of irrigation, fertilizer and other potentially harmful and costly cropping inputs (**a suitable means of establishing the spatial-variability of soil salinity at the field scale**), and (vii) a methodology for including soil salinity in the determination of plant-available soil water and for guiding irrigation management **(a suitable means for scheduling and controlling irrigations under saline conditions**).

From measurements of the levels and distributions of soil salinity within the root zones of individual fields, one can determine whether, or not, salinity is within acceptable limits for crop production. One can also infer from these measurements whether, or not, leaching and drainage are adequate anywhere in a field, since soil salinity is a tracer of the net processes of infiltration, leaching, evapotranspiration and drainage. Thus, a more appropriate and practical approach for assessing the adequacy of salinity control than either the L_r or SBI approaches is the acquisition of periodic, detailed information of soil salinity levels and distributions within the root zones of representative individual fields of the project. The same data can also be used for delineating the sources of salt-loading in fields and irrigated landscapes, as well as for mapping the distribution and extent of drainage problem areas, both at the project and field scales. The author refers to the above described approach as "salinity assessment" and advocates its use to diagnose, inventory and monitor soil salinity, as well as to evaluate the appropriateness of leaching and drainage and to guide management practices.

An assessment technology of the type described above begins with a practical methodology for measuring soil salinity in the field. This is complicated by the spatially variable and dynamic nature of soil salinity, which is caused by the effects and interactions of varying edaphic factors (soil permeability, water table depth, salinity of perched groundwater, topography, soil parent material, geohydrology), by management-induced factors (irrigation, drainage, tillage, cropping practices), as well as by climate-related factors (rainfall, amount and distribution, temperature, relative humidity, wind). Numerous samples (measurements) are needed to characterize just one field and the measurements often need to be updated as conditions change, or to determine if they are changing. When the need for extensive sampling requirements and repeated measurements are met, the expenditure of time and effort to characterize and monitor the salinity condition of a large area with conventional soil sampling and laboratory analysis procedures becomes impractical (as is shown later). Soil salinity is too variable and transient to be appraised using the numbers of samples that can be practically processed using conventional soil sampling and laboratory analysis procedures. Furthermore, the conventional procedures do not provide sufficient detailed spatial information to adequately characterize salinity conditions nor to determine its natural or management-related causes. A more rapid, field-measurement technology is needed. Additionally, this assessment technology should ascertain the spatial relations existing within extensive areal data sets. It should also provide a means for evaluating management effects and for proving changes or differences in an area salinity condition over time.

A system of the type advocated above has been developed. It consists of mobile instrumental techniques for rapidly measuring bulk soil electrical conductivity (EC_a) directly in the field as a function of spatial position on the landscape, procedures and software for inferring salinity from EC_a, computer-assisted mapping techniques capable of associating and analysing large spatial databases, and appropriate spatial statistics to infer salinity distributions in root zones and to detect changes in salinity over space and time. It will be described in some detail in this report. The complementary use of geographic information systems and remote sensing technology to facilitate the determination of the underlying causes of the observed salinity conditions would extend the utility of this system. The additional use of solute transport models utilizing the spatial data provided by the assessment system, as a basis to predict the consequences of alternative management practices, would extend its utility even more.

This report reviews the various electrical conductivity methods for determining soil salinity, for monitoring it and for mapping it, along with methodology for establishing the locations of measurement sites. Advantages and limitations of the alternative methods are discussed, including their relative costs; practical integrated mobile-systems for measurement/monitoring/ mapping applications are also described. Examples of the utility of the various methods are given for mapping and monitoring soil salinity, for diagnosing saline seeps, for evaluating the adequacy and appropriateness of irrigation and drainage management, for scheduling and controlling irrigations, for determining the leaching needed to reclaim saline soils, and for locating areal sources of over-irrigation and salt-loading. For earlier reports on instrumental field methods of soil salinity measurement and assessment see Rhoades (1976, 1978, 1984 and 1990a, b, 1992a, 1993, 1996a), Rhoades and Corwin (1984, 1990b), Rhoades and Miyamoto, (1990), Rhoades and Oster (1986) and Corwin and Rhoades (1990).

Chapter 2

Determination of soil salinity from aqueous electrical conductivity

The term salinity refers to the presence of the major dissolved inorganic solutes (essentially Na^+, Mg^{++}, Ca^{++}, K^+, Cl^-, $SO_4^=$, HCO_3^-, NO_3^- and $CO_3^=$) in aqueous samples. As applied to soils, it refers to the soluble plus readily dissolvable salts in the soil or, operationally, in an aqueous extract of a soil sample. Salinity is quantified in terms of the total concentration of such soluble salts, or more practically, in terms of the electrical conductivity of the solution, because the two are closely related (US Salinity Laboratory Staff, 1954).

PRINCIPLES OF AQUEOUS ELECTRICAL CONDUCTIVITY

Electrical conductivity (EC) is a numerical expression of the inherent ability of a medium to carry an electric current. Because the EC and total salt concentration of an aqueous solution are closely related, EC is commonly used as an expression of the total dissolved salt concentration of an aqueous sample, even though it is also affected by the temperature of the sample and by the mobility, valences and relative concentrations of the individual ions comprising the solution (water itself is a very poor conductor of electricity). Furthermore, not all dissolved solutes exist as charged-species; some are non-ionic and some of the ions combine to form ion-pairs which are less charged (they may even be neutral) and, thus, contribute proportionately less to electrical conduction than when fully dissociated.

The determination of EC generally involves the physical measurement of the materials' electrical resistance (R), which is expressed in ohms. The resistance of a conducting material (such as a saline solution) is inversely proportional to its cross-sectional area (A) and directly proportional to its length (L). Therefore, the magnitude of the measured electrical resistance depends on the dimensions of the conductivity cell used to contain the sample and of the electrodes. Specific resistance (R_S) is the resistance of a cube of the sample 1 cm on edge. Practical cells are not of this dimension and measure only a given fraction of the specific resistance; this fraction is the cell constant ($K = R/R_S$).

The reciprocal of resistance is conductance (C). It is expressed in reciprocal ohms, i.e., mhos. When the cell constant is applied, the measured conductance is converted to specific conductance (i.e., the reciprocal of the specific resistance) at the temperature of measurement. Often, and herein, specific conductance is referred to as electrical conductivity, EC:

$$EC = 1 / R_S = K / R. \qquad [1]$$

Electrical conductivity has been customarily reported in micro-mhos per centimetre (μmho/cm), or in milli-mhos per centimetre (mmho/cm). In the International System of Units (SI), the reciprocal of the ohm is the siemen (S) and, in this system, electrical conductivity is reported as siemens per metre (S/m), or as decisiemens per metre (dS/m). One dS/m is equivalent to one mmho/cm.

Electrolytic conductivity (unlike metallic conductivity) increases at a rate of approximately 1.9% per degree centigrade increase in temperature. Therefore, EC needs to be expressed at a reference temperature for purposes of comparison and accurate salinity expression; 25° C is most commonly used in this regard. The best way to correct for the temperature effect on conductivity is to maintain the temperature of the sample and cell at 25° ± 0.5°C while EC is being measured. The next best way is to make multiple determinations of sample EC at various temperatures both above and below 25° C, then to plot these readings and interpolate the EC at 25° C from the smoothed curve drawn through the data-pairs. For practical purposes of agricultural salinity appraisal, EC is measured at one known temperature other than 25° C and then adjusted to this latter reference using an appropriate temperature-coefficient (f_t). These coefficients vary for different salt solutions but are usually based on sodium chloride solutions, since their temperature coefficients closely approximate those of most salt-affected surface, ground, and soil waters. Another limitation in the use of temperature coefficients to adjust EC readings to 25° C is that they vary somewhat with solute concentration. The lower the concentration, the higher the coefficient, due to the effect that temperature has upon the dissociation of water. However, for practical needs, this latter limitation may be ignored and the value of f_t may be assumed to be single-valued. It may be estimated as:

$$f_t = 1 - 0.20346\,(T) + 0.03822\,(T^2) - 0.00555\,(T^3), \qquad [2]$$

where T = [temperature in degrees Celsius - 25] /10. This relation was derived from data given in Table 15 of Handbook 60 (US Salinity Laboratory Staff, 1954). In turn, the electrical conductivity at 25° C, EC_{25}, is calculated as:

$$EC_{25} = f_t * EC_t\,, \qquad [3]$$

where EC_t is the EC at the measured temperature t.

The above approach and f_t - temperature relations have been routinely used to reference soil electrical conductivity values (Rhoades, 1976), as well as solution/extract conductivities. The applicability of these f_t factors were tested for their appropriateness in this regard and concluded to be appropriate by McKenzie, *et al.* (1989), Johnston (1994), and Heimovaara (1995).

Because of differences in the equivalent weights, equivalent conductivities, and variations in the proportions of the various solutes found in soil extracts and water samples, the relationships between EC and total solute concentration and osmotic potential are only approximate. However, they are still quite useful. These relationships are as follows: total cation (or anion) concentration, mmoles charge/litre ≅10 x EC_{25}, in dS/m; total dissolved solids, mg/litre ≅ 640 x EC_{25}, in dS/m; and osmotic potential, M Pa at 25° C ≅ 0.04 x EC_{25}, in dS/m.

SOIL WATER SALINITY

Theoretically, the electrical conductivity of the soil solution (EC_w) is a better index of soil salinity than is the traditional index (EC_e). This is so because the plant roots actually experience the soil solution; they extract their nutrients from it, absorb other solutes from it and they consume this water through the process of transpiration. However, EC_W has not been widely used as a means for measuring or expressing soil salinity for several reasons. Firstly, it is not single-valued; it varies over the irrigation cycle as the soil water content changes (Rhoades, 1978). Thus, EC_W does not lend itself to simple classifications or standards unless it is referenced to a specific water content, such as field capacity. Secondly, and probably most importantly, EC_W has not been widely adopted for routine appraisals of soil salinity because methods for obtaining soil water samples are not practical at typical field water contents.

Samples of soil solutions may be obtained from soil samples in the laboratory by means of displacement, compaction, centrifugation, molecular adsorption and vacuum- or pressure-extraction methods. The latter methods are described by Richards (1941); displacement methods by Adams (1974); combination displacement/centrifugation methods by Gillman (1976), Mubarak and Olsen (1976, 1977) and Elkhatib *et al.* (1986); a combination vacuum/displacement method by Wolt and Graveel (1986); a simple field-pressure filtration method by Ross and Bartlett (1990); adsorption techniques by Davies and Davies (1963), Yamasaki and Kishita (1972), Gillman (1976), Dao and Lavy (1978), Kinniburgh and Miles (1983) and Elkhatib *et al.* (1987). Comparisons of the various methods have been made by Adams *et al.* (1980); Kittrick (1983); Wolt and Graveel (1986); Menzies and Bell (1988) and Ross and Bartlett (1990).

Two means of measuring EC_W in undisturbed soils exist. One is collect a sample of soil water using an in-situ extractor and then to measure its EC; the second is to measure EC_W "directly" in the soil using in-situ, imbibition - type "salinity sensors".

A typical vacuum extractor system used to collect soil water samples in the field is shown in Figure 1. This suction-method, first proposed by Briggs and McCall (1904), is useful for extracting water from the soil when the soil-water suction is less than about 0.1 M Pa. Although the available range of soil moisture for crops extends to 1.5 M Pa of soil suction, most water uptake by plants takes place within the zero to 0.1 M Pa range. Therefore, the suction method is applicable for many salinity-monitoring needs. While different extraction devices have been used, the most commonly used is the porous ceramic cup. Early vintage extractor construction and performance are described in a bibliography assembled by Kohnke *et al.* (1940). Reeve and Doering (1965) described the more modern equipment and procedures for its use in detail. These procedures have been used in field experiments with good success for salinity appraisal purposes. Wagner (1965) used similar devices to estimate nitrate losses in soil percolate. Other, improved and specialized, versions have since been developed for various purposes, including the following: a miniature sampler which eliminates sample transfer in the field (Harris and Hansen, 1975), samplers which shut off automatically when the desired volume of sample is collected (Chow, 1977), samplers which function at depths greater than the suction lift of water (Parizek and Lane, 1970; Wood, 1973) and samplers which minimize "degassing" effects on solution composition (Suarez, 1986, 1987). Soil water has also been extracted using cellulose-acetate hollow fibres (Jackson *et al.* 1976; Levin and Jackson, 1977), which are thin-walled, semipermeable, and flexible. Claimed advantages include flexibility, small diameter, minimal chemical interaction of solutes with the tube matrix, and compositional results comparable with those from samples obtained from ceramic extraction cups. Collection "pan"-type collectors have also been used to collect soil percolate (Jordan, 1968). Additionally, large-scale vacuum extractors (15 cm wide by 3.29 m long) have been built and used to assess deep percolation losses and chemical composition of soil water (Duke and Haise, 1973). Ceramic "points", which absorb water upon insertion into the soil, have also been used to sample soil water with some

FIGURE 1
Diagram of vacuum extractor apparatus for sampling soil water (after Rhoades and Oster, 1986)

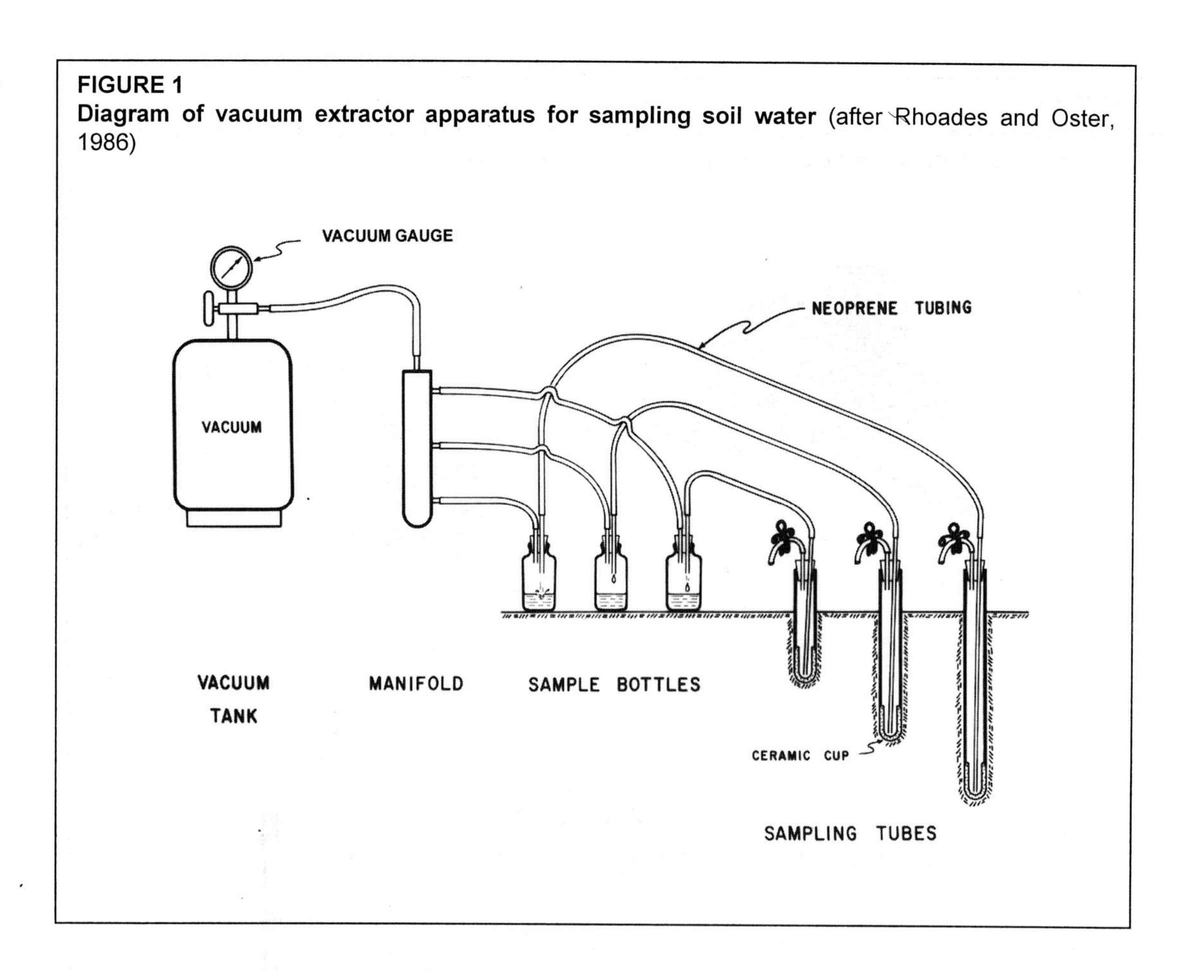

success (Shimshi, 1966). However, only very small samples are obtained with these "points" and there are potential errors due to vapor transfer and chromatographic separation. Tadros and McGarity (1976) have analogously used an absorbent sponge material.

Various errors in sampling soil water can occur with the use of any of the above types of extractors. Included are factors related to sorption, leaching, diffusion, and sieving by the cup wall; also to sampler intake rate, plugging, and sampler size. Nielsen *et al.* (1973), Biggar and Nielsen (1976), and van De Pol *et al.* (1977), used soil water extractors to determine salt flux in fields and have demonstrated that field variability in this regard is very large. They concluded that soil water samples being "point samples" can provide only indications of relative changes in the amount of solute flux, but not of quantitative amounts, unless the frequency distribution of such measurements is established. Because the composition and concentration of soil water is not homogeneous through its entire mass; water drained from large pores at low suctions (such as that collected by vacuum extractors) may have compositions very different from that extracted from micropores. A point source of suction, such as a porous cup, samples a sphere of different-sized pores dependent upon distance from the point, the amount of applied suction, the hydraulic conductivity of the medium, and the soil water content. Although vacuum extractors are versatile, easily usable and provide for *in situ* sampling of soil water, they have, as evident from the above discussion, limitations. For more discussion of the different suction-type samplers and other methods for sampling soil solution and various errors associated with them see the reviews by Rhoades (1978, 1979a), Rhoades and Oster (1986), Litaor (1988) and Grossman and Udluft (1991).

FIGURE 2
(A) Imbibition-type salinity sensor with spring, housing and pin in disassembly; (B) schematic of internal elements of salinity sensor (after Rhoades and Oster, 1986)

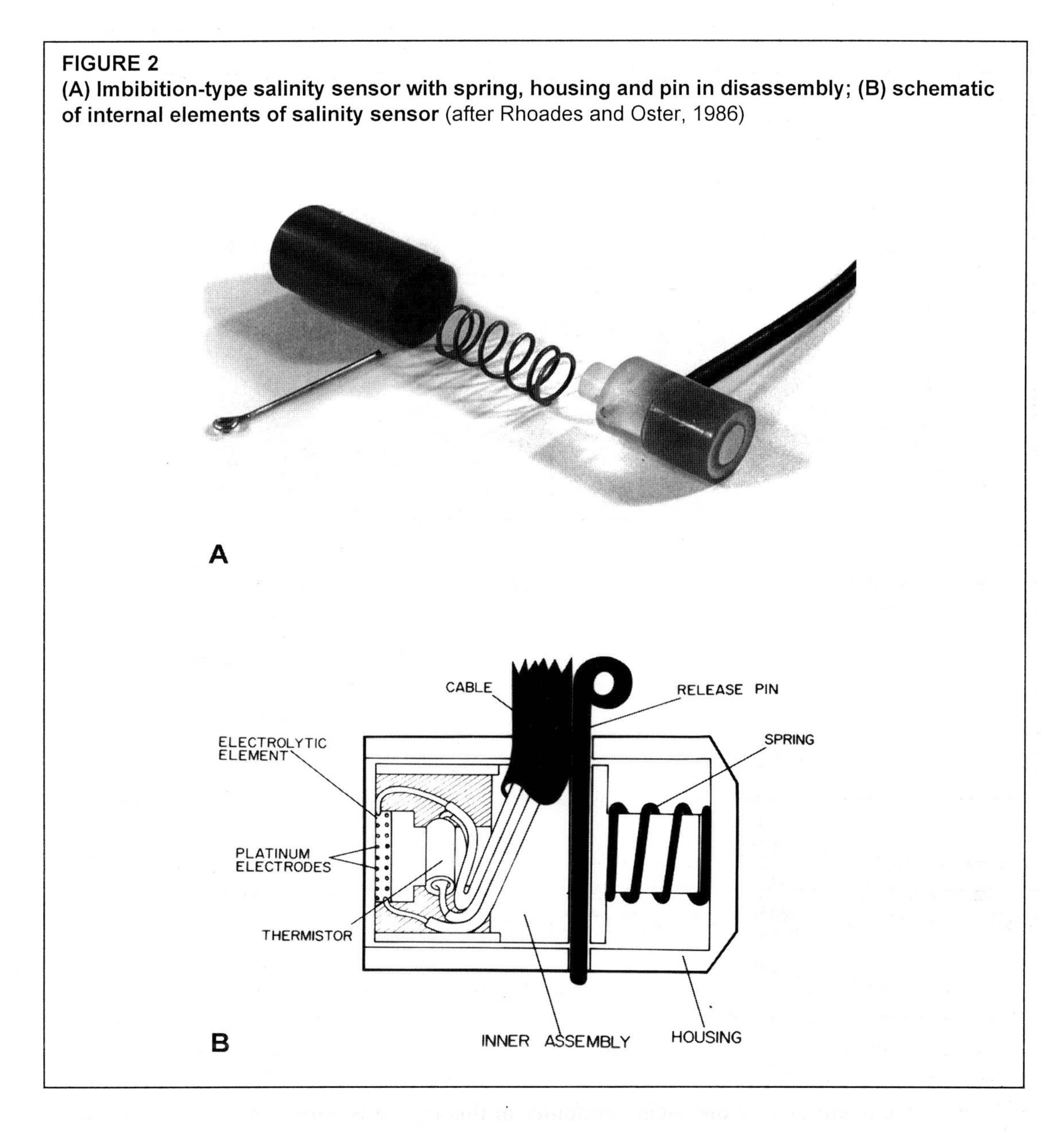

When the total concentration of salts in the soil water is sufficient information, i.e., when specific solute analyses are not needed, in-situ devices capable of directly measuring EC_W may be used advantageously. Kemper (1959) developed the first *in situ* salinity sensor. It consisted of electrodes imbedded in porous ceramic to measure the electrical conductivity (EC) of the solution within the "ceramic cell". When placed in soil, these devices imbibe water which, in time, comes to diffusional equilibrium with the soil water. Richards (1966) improved the design of the soil salinity sensor to shorten its response time and to eliminate external electrical current paths. This unit is now produced commercially. In this unit (Figure 2), the salinity sensitive element is an approximately 1-mm-thick ceramic plate which contains platinum screen electrodes on opposite sides. This gives a short diffusion path and thus lowers response time. Another feature of the design is a preloaded spring. After the salinity sensor is placed in the soil, the spring is released to ensure good contact of the ceramic plate with the soil. Athermistor is incorporated in the sensor so that the EC may be adjusted for temperature effects. A commercially available meter developed for these sensors is shown in Figure 3. An oscillator

FIGURE 3
Commercial meter and salinity sensor showing ceramic disc in which platinum electrodes are embedded (after Rhoades, 1993)

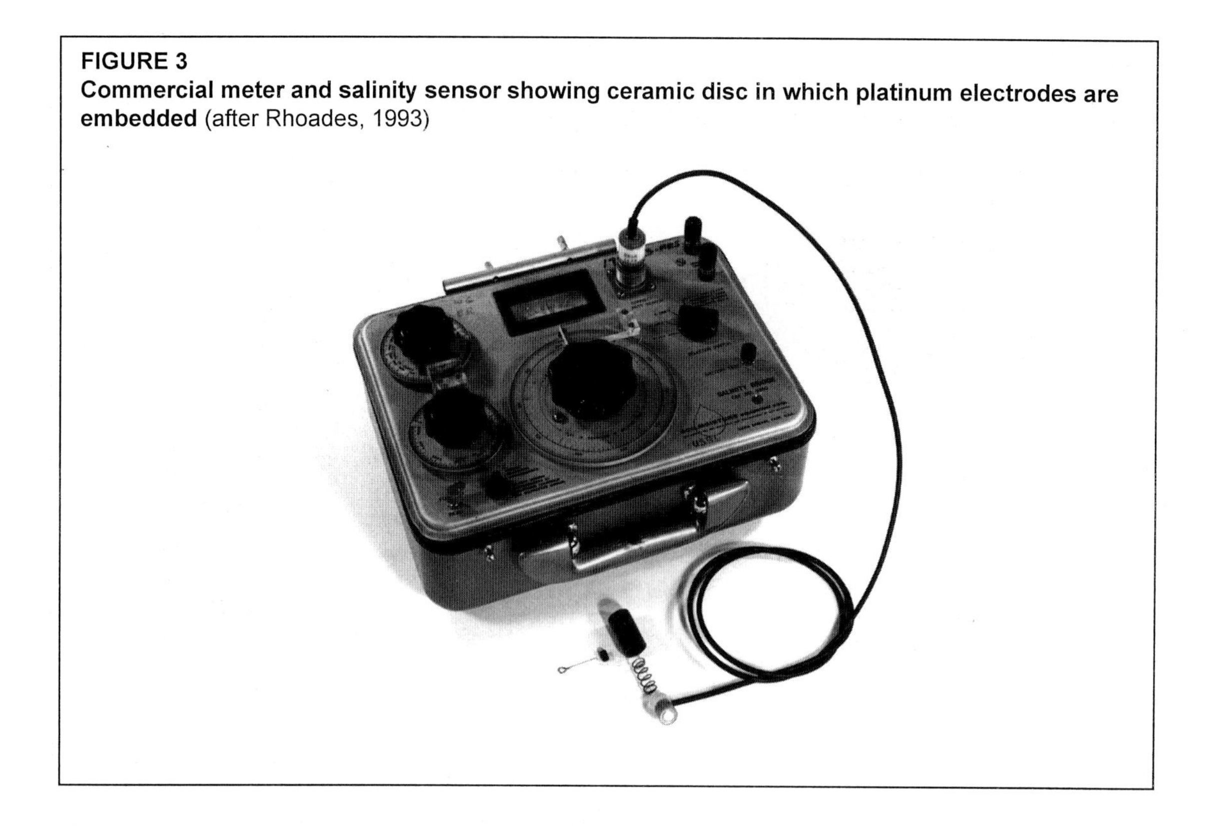

circuit system has been developed for automated salinity sensor measurements and data logging (Austin and Oster, 1973). This permits linear readings to be obtained with lead lengths of up to several hundred meters.

Salinity sensors have been mostly used in agricultural research, where continuous monitoring of soil salinity in soil columns, lysimeters, and field experiments is required (Oster and Ingvalson, 1967; Rhoades, 1972; Oster *et al.* 1973; Oster *et al.* 1976; Ingvalson *et al.* 1970). The accuracy of the commercial ceramic sensor under such conditions has been found to be ± 0.5 dS/m (Oster and Ingvalson, 1967). Reliability was determined by removing the sensors from field and lysimeter experiments after 3-to-5-years of continuous operation and comparing their calibrations relative to original ones (Oster and Willardson, 1971; Wood, 1978). About 68 percent of the tested sensors had calibrations within 14 percent of the original calibrations after five years. Shifts varied in direction and magnitude, and some complete failures occurred.

Response times of the commercial salinity sensors have been evaluated in field situations (Wesseling and Oster, 1973; Wood, 1978). In the matric potential range of -0.05 to -0.15 M Pa, 90 percent of the response of these sensors to a step change in salinity will occur within 2 to 5 days. At lower matric potentials, response times are longer. Thus, it may be concluded that salinity sensors are not well suited for measuring short-term changes in salinity because of their relatively long response time of at least several days. Desaturation of the ceramic occurs at matric potentials more negative than -0.2 M Pa, significantly reducing the conductance of the ceramic salinity sensor (Ingvalson *et al.* 1970). Hence, this type of sensor is not accurate in "dry" soils. Salinity sensors constructed of porous glass have been developed which remain saturated with soil water to 2 M Pa matric potentials (Enfield and Evans, 1969), but they are fragile and not available commercially.

FIGURE 4
Variations in soil water electrical conductivity (EC_W) and soil water tension in the root zone of an alfalfa crop during the spring of the year (after Rhoades, 1972)

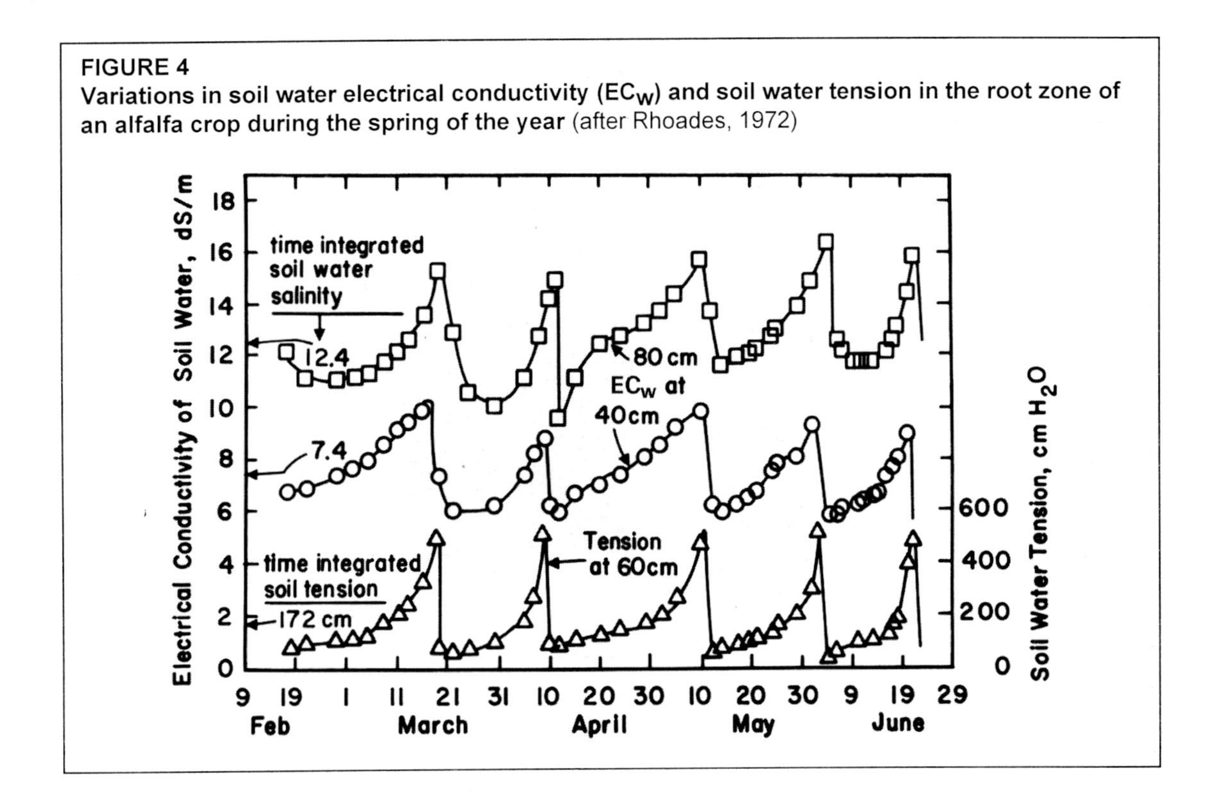

Soil disturbance during installation can result in errors associated with modified water infiltration in the back-filled hole used to install salinity sensors. Special precautions during their installation must be taken to avoid this.

While, obviously, also with limitations, salinity sensors may be used advantageously for continuously monitoring electrical conductivity of soil water at selected depths over relatively long periods of time, as illustrated in Figure 4. They are not well suited for measuring short-term changes of salinity, especially in "dry" soils. Many units may be needed because of their small sampling volume, and the substantial heterogeneity of soils, in order to characterize the actual conditions existing in irrigated soils. These numbers can be minimized if the sensors are primarily used to follow changing salinity status at a specific location over time. They are simple in principle, easily read, and sufficiently accurate for intermediate-term salinity monitoring purposes. They must be individually calibrated; these calibrations may change with time. They are, of course, not practical for mapping purposes for obvious reasons.

SOIL EXTRACT SALINITY

Because present methods of obtaining soil water samples at typical field water contents are not very practical, aqueous extracts of the soil samples have traditionally been made in the laboratory at higher-than-normal water contents for routine soil salinity diagnosis and characterization purposes. Since the absolute and relative amounts of the various solutes are influenced by the water/soil ratio at which the extract is made (Reitemeier, 1946), the water/soil ratio used to obtain the extract should be standardized to obtain results that can be applied and interpreted reasonably generally. As stated earlier, soil salinity is most generally defined and measured on aqueous extracts of so-called, saturated soil-pastes (US Salinity Laboratory Staff, 1954). The water content of saturated soil-pastes (the so-called saturation percentage, SP), as well as the water/soil ratio, varies with soil texture. It is related in a reasonably general and predictable way to soil-water contents under field conditions. For these same reasons, crop tolerance to salinity is also most generally expressed in terms of the electrical conductivity of the saturation-extract (EC_e, Maas and Hoffman, 1977; Maas, 1986, 1990). Herein, the term saturated soil-paste extract is often used in place of saturation-extract; the two terms are synonymous.

Estimates made of the EC_w from EC_e and the ratio of their water contents will usually be excessively high. This is because salts will often be present in the saturation-extract that would not be under actual field conditions. Additionally, salts contained within the fine pores of aggregates will contribute to the EC_e value, though it is doubtful that significant amounts of such salts are absorbed by plant roots or affect the availability of the majority of the water extracted by the plant (which is primarily that present in the larger pores).

FIGURE 5
(A) Saturated soil-paste in mixing container and in drying/weighing container; (B) saturated soil-paste being extracted by vacuum; and (C) collection of saturated-paste extract

EC_e is typically determined as follows. A saturated soil-paste is prepared by adding distilled water to a sample of air dry soil (200 to 400 g) while stirring and then allowing the mixture to stand for at least several hours (but often overnight) to permit the soil to fully imbibe the water and the readily soluble salts to fully dissolve, so as to achieve a uniformly saturated and equilibrated soil-water paste (see Figure 5A). At this latter point, which is sufficiently reproducible for practical purposes, the soil paste glistens as it reflects light, flows slightly when the container is tipped, slides freely and cleanly off a spatula, and consolidates easily when the container is tapped or jarred after a trench is formed in the paste with the broad side of the spatula. The extract of this saturation-paste is usually obtained by suction using a funnel and filter paper (see Figure 5B and 5C). The EC and temperature of this extract are then measured using standard conductance meters/cells and thermometers, respectively (see Figure 6); the EC_{25} value of this extract is calculated from Equation [3] to give EC_e.

Once soil extract samples are obtained, laboratory chemical analyses can be carried out to determine, in addition to the electrical conductivity of the extract (EC_e), the concentrations of the individual solutes, i.e., Na^+, Ca^{++}, Mg^{++}, K^+, Cl-, $SO_4^=$, HCO_3-, $CO_3^=$, NO_3-, etc. Methods for such analyses are given elsewhere (Rhoades, 1982; Soil Science Society of America, 1996). More details about the methods for measuring the electrical conductivity and total dissolved solids contents of aqueous samples and extracts are given in Rhoades (1982, 1993, 1996a); for a good discussion of some of the operational factors influencing the procedure see Shaw (1994).

FIGURE 6
Measurement of the electrical conductivity of a saturated-paste extract (EC_e) using a laboratory micro-conductivity cell

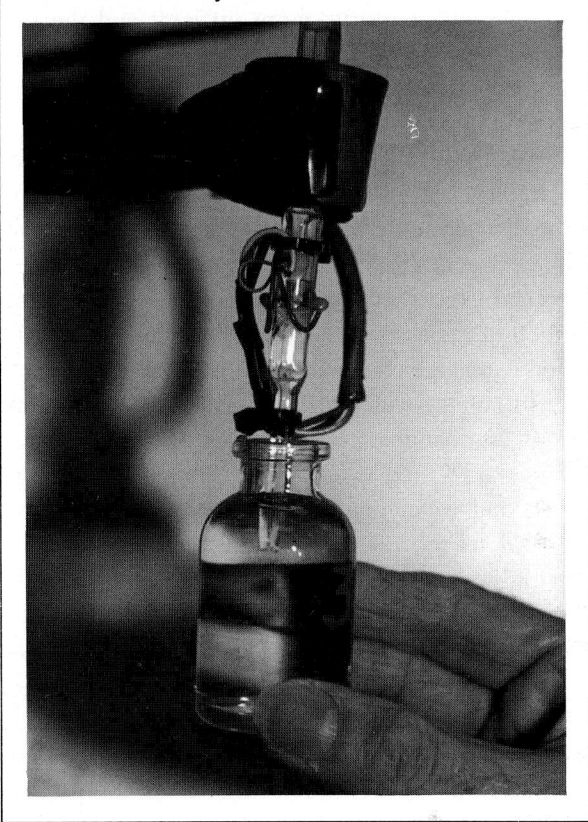

Though the above-described procedure for making a saturated soil-paste is somewhat subjective, diagnoses are not compromised by the normal variations experienced in it. Yet this subjectivity seems to be a concern to some people (Shaw, 1994). To eliminate some of the subjectivity of the saturation extract method, Longenecker and Lylerly (1964) proposed wetting the sample by capillarity using a "saturation table". Beatty and Loveday (1974) andLoveday (1972) advocated predetermining the amount of water at saturation on a separate soil sample using a similar capillary wetting technique and then adding this amount to all other samples of the same soil. Allison (1973) recommended slowly adding soil to water, rather than water to soil, when making pastes to speed wetting of the soil and preparation of the saturated-paste condition. All of these modifications offer advantages over the standard procedure under certain situations, but all but the last one slow the procedure considerably without significantly enhancing the diagnostic value of the result.

Other extraction ratios, such as 1:1, 1:5, etc., are easier to use than that of the saturation paste but they are less well-related to meaningful soil properties and are more subject to errors resulting from peptization, hydrolysis, cation exchange, and mineral dissolution. Sonnevelt and van den Ende (1971) recommended a 1:2 **volume** extract. This method is a compromise between the saturation-paste extract and the higher-dilution "weight" extracts. The water contents of the 1:2 volume "pastes" of sandy and clayey soils are higher and lower, respectively, relative to the saturation-paste extract. For purposes of monitoring, when relative changes are of more concern than the absolute solute concentration(s), these quicker, simpler methods of "fixed-extraction-ratios" may be used to advantage in place of the saturation extract. Of course, the relations given in Handbook 60 to predict exchangeable sodium percentage from the sodium adsorption ratio apply only to the saturation-paste extract, as do most of the other indices/criteria/ standards used to express/interpret soil salinity/sodicity/toxicity and plant response (salt-tolerance, plant-growth data) from soil analyses. Criteria for evaluating salinity and sodicity effects on soils and crops have been developed for some of the higher-dilution extracts, but the greater errors and lack of uniformity they create in this regard makes the extrapolation of results more difficult and the literature more confusing (Rengasamy, 1997). Consistency and uniformity of methodology and criteria/standards should be sought whenever possible to facilitate interpretation of results and their general applicability.

Because of the numerous interacting effects of the following: mineral dissolution/ precipitation, cation exchange, ion-pair formation, negative adsorption, time of equilibration, amount of grinding and drying, presence or absence of suspended minerals and organic matter in the extract, microbial production of CO_2 during equilibration, etc., a computer deterministic-chemical model is required, along with the determination of the ionic-composition of the extract, of the associated cation exchange composition, of the cation-exchange-capacity and the cation exchange coefficients, in order to accurately calculate the EC of a solution in association with soil as the water/soil ratio is altered (Paul *et al.*, 1966). The assumption of conservation of mass with change of water content during a change in water content is not sufficiently valid to permit the EC at a second water content to be accurately calculated as the product of the EC at the second water content and the ratio of the two water contents. Since the computer-model approach is too demanding and the second ratio approach too simplistic, various empirical relations have been developed to estimate EC_e from EC values measured on higher-dilution extracts. Shaw (1994) has reviewed most of these methods and concluded that they are very location-specific and can not be extrapolated reliably elsewhere. He developed an improved, more generally applicable relation to estimate EC_e from the EC of a 1:5, soil:water extract, but it requires an analysis of the extract for chloride concentration and the determination of the air dry moisture content of the soil sample used to make the 1:5 extract. This method and all of the others like it will be more accurate for solutions dominated by chloride salts; very substantial errors may occur with soils containing gypsum, especially if they are also sodic (Adiku *et al.*, 1992). One must question whether any of these "conversion" methods save sufficient time and effort to make them worthwhile, especially considering the uncertainty in their resulting estimates of EC_e. A faster and more accurate method for estimating EC_e, based on simple measurements of the volume-weight and EC of the saturated-paste itself (rather than of its extract), is described in the following chapter. This method eliminates much of the work involved in measuring EC_e and SP using conventional methods (the latter involves oven-drying for 24 hours) without the loss of accuracy that occurs in estimating it from EC measurements made on extracts obtained at higher dilutions. Additionally, one obtains the added SP information with this method, which is valuable as an estimator of many soil properties including texture, water-holding capacity and cation-exchange capacity.

Chapter 3

Determination of soil salinity from soil-paste and bulk soil electrical conductivity

The methods of soil salinity determination already described are not well suited for use in the field nor for intensive-mapping and monitoring applications because they require the collection of soil samples and their aqueous extracts. Thus, they are relatively slow and expensive to carry out (see later discussions and Chapter 5). For this reason, more practical field methods have been sought and developed. One such method eliminates the need for aqueous extractions, though it still requires the collection of soil samples and the making of saturated soil-pastes. Another still more practical method is based on direct measurements of bulk soil electrical conductivity (EC_a) made upon undisturbed soils using geophysical-type sensors; this methodology is especially well suited for intensive mapping and monitoring applications. These two methods are described in this section.

PRINCIPLES OF SOIL AND SOIL-PASTE ELECTRICAL CONDUCTIVITIES

A model of the electrical conductivity of mixed soil/water systems that has been shown to be very useful and generally applicable for purposes of salinity appraisal is illustrated in Figure 7. This model supersedes the earlier model of Rhoades *et al.* (1976). It assumes that the electrical conductivity of a soil containing dissolved electrolytes (salts) in the soil "solution" can be represented by conductance via the following three pathways (or elements) acting in parallel: (1) conductance through alternating layers of soil particles and the soil solution that envelopes and separates these particles (a solid-liquid, series-coupled element), (2) conductance through continuous soil solution pathways (a liquid element), and (3) conductance through or along the surfaces of soil particles in direct and continuous contact with one another (a solid element).

Because most soil minerals are insulators, electrical conduction in sufficiently moist soils is primarily via the electrolytes (salts) contained in the water occupying the larger pores. The contribution of the solid phase to electrical conduction in moist soils, the so-called surface conduction, is primarily via the exchangeable cations associated with the clay minerals (though the latter are actually present in the aqueous phase). Surface conductance is generally smaller than that of the pore-solution because the former electrolytes are more limited in their amounts and mobilities. The magnitude of surface conduction is assumed (and tests have confirmed this) in the soil electrical conductivity model to be, for practical purposes, independent of the dissolved salts and essentially constant for any given soil (Rhoades *et al.*, 1976; Shainberg *et al.*, 1980; Bottraud and Rhoades, 1985a). The surface conductance is also assumed to be coupled in series with the electrolyte present in the water films associated with the solid

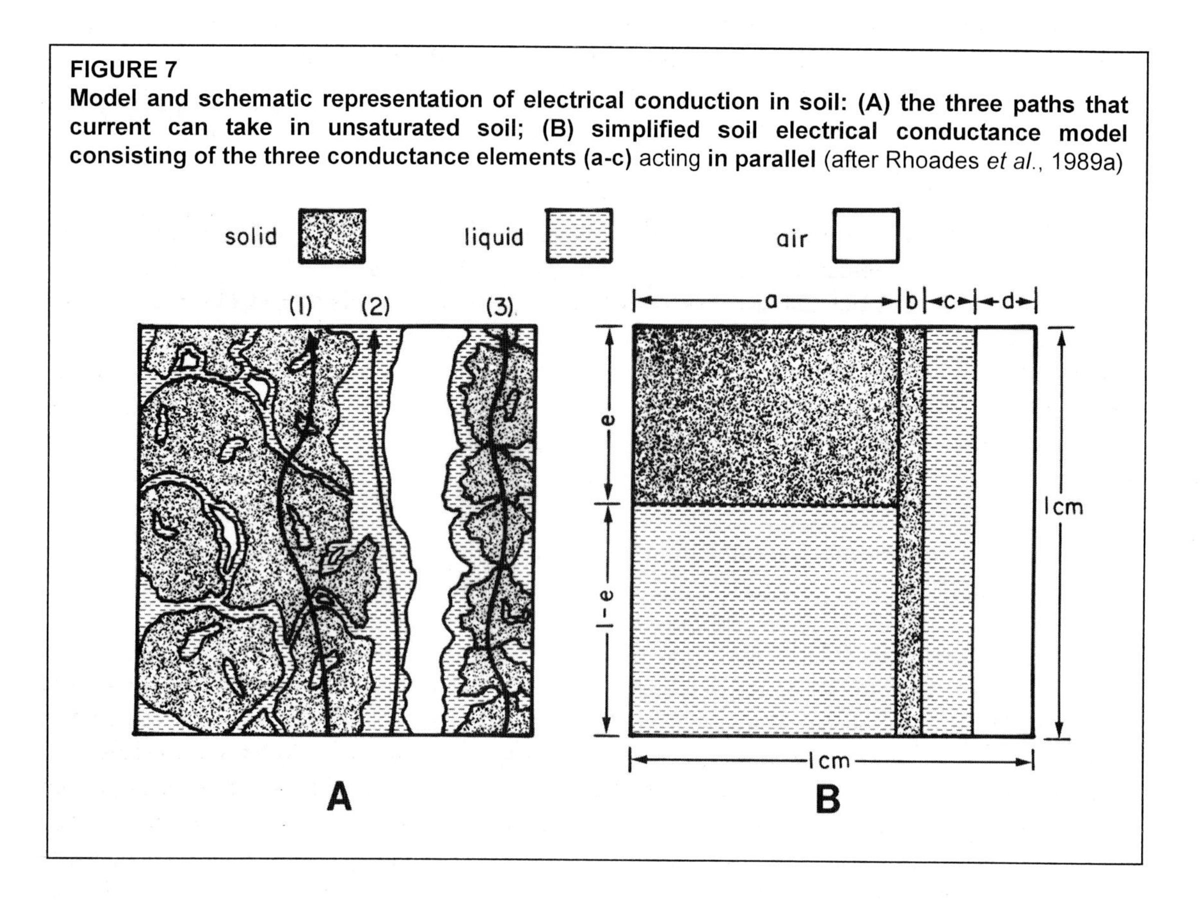

FIGURE 7
Model and schematic representation of electrical conduction in soil: (A) the three paths that current can take in unsaturated soil; (B) simplified soil electrical conductance model consisting of the three conductance elements (a-c) acting in parallel (after Rhoades *et al.*, 1989a)

surfaces and in the small water-filled pores which serve as "links" between adjacent particles and aggregates to provide a secondary pathway for current flow in moist soils. This pathway is modelled as acting in parallel with the primary, continuous pathway (salt-solution contained in the large water-filled pores). The solid element pathway may exist in soils with indurated layers. In such layers, conductance could occur through, or along the surfaces of, the soil particles, which are in direct and continuous contact with one another (a solid element). The relative flow of current in the three pathways depends upon the volumetric contents and solute concentrations of the water in the two different categories of pores and on the volumetric contents and magnitudes of the surface-conduction and of the indurated solid-phase. This model is mathematically represented by Equation [4]:

$$EC_a = \left[\frac{(\Theta_{ss} + \Theta_{ws})^2 \; EC_{ws} \; EC_{ss}}{(\Theta_{ss}) \; EC_{ws} + (\Theta_{ws}) \; EC_{s}} \right] + \Theta_{sc} \; EC_{sc} + \Theta_{wc} \; EC_{wc} \; , \qquad [4]$$

where EC_a is the electrical conductivity of the bulk soil; Θ_{ws} and Θ_{wc} are the volumetric soil water contents in the series-coupled pathway (small pores) and in the continuous liquid pathway (large pores), respectively; Θ_{ss} and Θ_{sc} are the volumetric contents of the surface-conductance and indurated solid phases of the soil, respectively; EC_{ws} and EC_{wc} are the specific electrical conductivities of the soil water that are in series-coupling with the solid particles and in the continuous conductance element, respectively, and EC_{ss} and EC_{sc} are the electrical conductivities of the surface-conductance and indurated solid phases, respectively. The soil water in the continuous pathway is envisioned as the water occupying the larger pores, commonly referred to as "mobile" water. This water can be different in composition from that in the small pores and intra-ped pores, which is envisioned as the "immobile" water associated in the model with the series-coupled pathway. Ultimately, diffusion processes will cause EC_{ws} and EC_{wc} to be equal. However, when water is being added by irrigation or rain, or is being removed by drainage or evapotranspiration, equilibrium will not exist; consequently, EC_{ws} and EC_{wc} may be different during these periods.

The second term of the second member in Equation [4], i.e. Θ_{sc} EC_{sc}, usually may be dropped. This is so, apparently, because soil structure simply does not allow for enough direct particle-to-particle contact between aggregate units in typical agricultural soils to provide a continuous solid-phase pathway for electrical current flow. This latter potential, pathway is disrupted by water films surrounding the particles and peds or by void spaces within the matrix that are filled with either liquid or air. Experimental data show it to be negligible (Rhoades *et al.*, 1976, 1990a). Thus for all but soils with indurated layers, Equation [4] may be simplified to the following two-pathway model (Rhoades *et al.* 1989a):

$$EC_a = \left[\frac{(\Theta_s + \Theta_{ws})^2 \; EC_{ws} \; EC_s}{(\Theta_s) \; EC_{ws} + (\Theta_{ws}) \; EC_s} \right] + (\Theta_w - \Theta_{ws}) \; EC_{wc} \, , \qquad [5]$$

where (Θ_w - Θ_{ws}) is substituted for Θ_{wc} , θ_w is the total volumetric soil water content, and EC_s is the surface conductance of soils without indurated layers. This equation has been shown to be generally applicable to arid-land mineral soils of the Southwestern United States (Rhoades *et al.*, 1989a, 1990a). There is no reason to believe that it is not equally applicable to similar arid-land soils found elsewhere in the world. However, the model has not been tested on soils containing high contents of gypsum, which may differ because gypsum particles may be more conductive than silicate mineral particles. This could result in higher values of EC_{ss} and EC_{sc}. This problem has not been observed by the author in gypsiferous US soils, but they do not contain as high of gypsum contents as occur in some parts of the Near East; no reports have been found indicating that others have observed this to be a significant problem.

For conditions of EC_{ws} greater than about 2-4 dS/m and for soils with typical values of EC_s (less than about 1.5 dS/m), the product (Θ_s * EC_{ws}) is so much larger than the product (Θ_{ws} * EC_s) that the latter product can be neglected; thus simplifying Equation [5] to the following one **for typical saline soils**:

$$EC_a = \left[\frac{(\Theta_s + \Theta_{ws})^2 \; EC_s}{(\Theta_s)} \right] + (\Theta_w - \Theta_{ws}) \; EC_{wc} \, . \qquad [6]$$

Equation [5] is the more generally applicable relation and must be used for soils with low values of EC_{ws}, i.e. non-saline soils, where the relation between EC_a and EC_{wc} is curvilinear at low levels of EC_{wc}. The first term of the second member of this equation determines the shape of the nonlinear portion of the EC_a-EC_{wc} curve. Over the remainder of the EC_{wc} range, EC_a and EC_{wc} are linearly related, with (Θ_w - Θ_{ws}) representing the slope of this relation. In contrast, the simplified relation expressed in Equation [6] should only be used for conditions of $EC_{wc} \geq$ about 2 dS/m ($EC_e \geq$ about 1-2 dS/m) and $EC_s \leq$ 1.5 dS/m, i.e., for typical saline soils. For such cases, the relation between EC_a and EC_{wc} expressed in Equation [6] is linear and proportional to (Θ_w - Θ_{ws}) beyond the threshold value of EC_{wc} and the y-intercept depends upon EC_s, Θ_s and Θ_{ws}. Since the ratio $[(\Theta_s + \Theta_{ws})^2 / \theta_s]$ is typically close to the value 1 (because Θ_s is typically about 0.5 and Θ_{ws} is less than or equal to 0.5 Θ_w , where Θ_w is typically about 0.4 or less); the intercept of Equation [6] is approximately equal to EC_s and may be symbolized as EC_s*. The earlier EC_a model of Rhoades, et al. (1976) is analogous to this limiting case version of Equation [6], as shown elsewhere (Rhoades *et al.* 1989a). This earlier model expressed the slope in terms of a tortuosity concept, but it is mathematically identical to that expressed in Equations [5] and [6] which supersede it. The major improvement is contained within the intercept term of Equation [5].

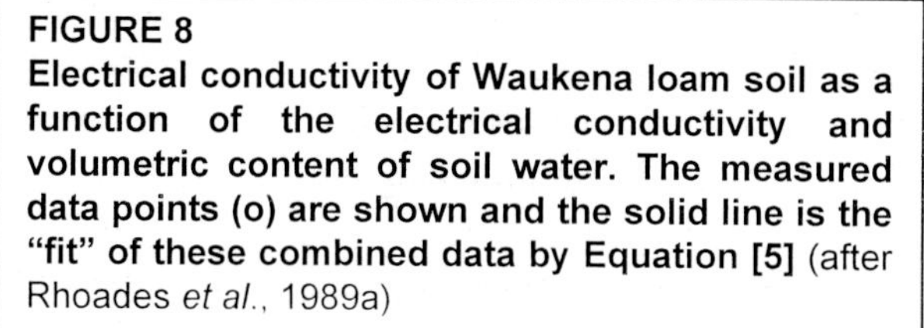

FIGURE 8
Electrical conductivity of Waukena loam soil as a function of the electrical conductivity and volumetric content of soil water. The measured data points (o) are shown and the solid line is the "fit" of these combined data by Equation [5] (after Rhoades *et al.*, 1989a)

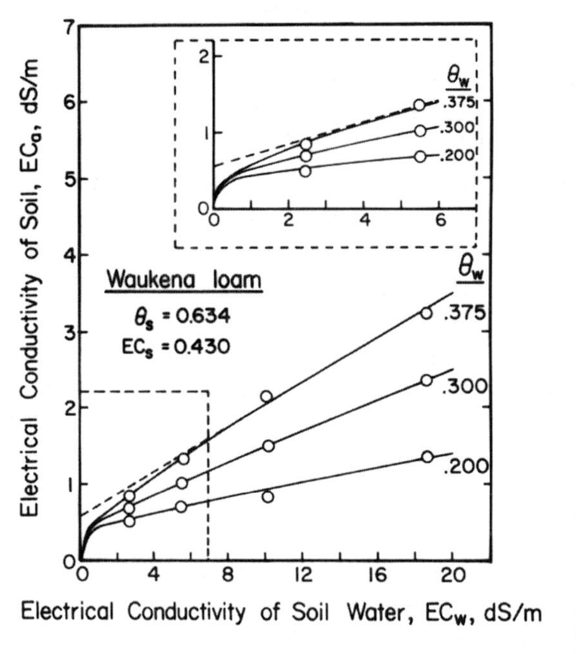

Data illustrating the appropriateness of the above described model and generalizations are shown in Figure 8 for Waukena loam soil. The solid line is that described by Equation [5], the dashed line is that described by Equation [6] for the one example water content (0.375, for purposes of illustration), and the circles represent experimental data. EC_w represents the EC of the equilibrating water or the water extracted from the soil by pressure filtration. Note that salts, as well as water, were removed during the pressure filtration of the soil that was used in this controlled-experiment to vary water content while keeping EC_w constant. The soil had been extensively leached with waters of different salinities (EC_w values), therefore EC_{wc} and EC_{ws} were essentially equal to EC_w under the conditions of this experiment. These data and the model relations also show that EC_w can be inferred from measurements of EC_a made at relatively low water content, but the ability to accurately do so decreases as Θ_w decreases. This is so because the required accuracy of measurement of EC_a becomes limiting as the $EC_a = f(EC_w)$ relation flattens at low values of θ_w (Rhoades *et al.*, 1976; Bottraud and Rhoades, 1985b). However, at very low values of Θ_w, it is not possible to determine EC_w (or EC_e) from EC_a at all. The value of the "threshold" water content is approximately 0.10. For more discussion and data about the threshold water content see Figure 6 and Table 3 in Rhoades *et al.* (1976). "Dry" soil measurements are to be avoided for the reason given above and for those which follow.

As explained above, the ability to accurately determine EC_w (or EC_e) from EC_a decreases as Θ_w decreases. For this reason, it is recommended that EC_a measurements be limited to moisture contents that are not less than about one-half of field-capacity water content. Most irrigated soils are kept above this level during the cropping season. As also stated above and elsewhere (Rhoades *et al.*, 1976), it is not possible to measure EC_a at very low values of Θ_w; nor is it possible to use measurements of EC_a to determine salinity under such conditions. This is so because there must be a continuous pathway for electrical flow through the soil in order to make the measurement. It should be noted that the soil-EC model assumes the presence of sufficient moisture to permit current flow to take place via the two pathways existing within the soil matrix (the water phase which is in continuous contact via the larger soil pores and the water films which envelope and bridge soil particles to form another continuous pathway). As explained above, the threshold value of Θ_w required to satisfy the above requirement is about 0.1, possibly more in sandy soils. This minimum limit will usually be met in all but the surface dry-mulch layer of irrigated soils during most of the irrigation season. However, it is another matter for dryland soils. Since dry soil is essentially an insulator, no useful information about salinity, or other soil properties for that matter, can be inferred from EC_a measurements made on such dry soils. Therefore, one should not include the dry surface mulch in samples used to calibrate EC_a - soil properties. EC_a measurements should only be made in dryland soils during the time of the year when they are sufficiently moist for the measurable-conduction of electricity. It is sufficiently important to repeat: **it is inappropriate to try to infer salinity from measurements of EC_a made on dry, or nearly dry, soil as it is to include salinity analyses of such soils in the data used to establish EC_a - EC_e calibrations**. This will be commented on later when the relative merits of the different sensors which can be used to measure EC_a, as well as the different methods of calibration, are discussed.

Since, $\Theta_s = \rho_b / \rho_s$, soil bulk density ($\rho_b$) and soil particle density (ρ_s) are two soil properties, besides salinity, that affect EC_a. The value of EC_a is also affected by clay content and type, since EC_s is primarily associated with the cation exchange capacity. Additionally, EC_a is expected to be affected by the pore size distribution and structure of the soil, since they influence the contents of "mobile" (Θ_{wc}) and "immobile" (Θ_{ws}) water. Likewise, prior events (i.e., irrigation, rainfall, evapotranspiration) and processes (i.e., diffusion) which influence the distributions of salt concentrations between the mobile and immobile phases (EC_{wc} and EC_{ws}, respectively) can affect EC_a. The sensitivity of EC_a to each of these factors can be determined from Equation [5] to the extent that they can be related to the parameters used in the model, for example EC_s = f (% clay, clay type), Θ_s and Θ_w = f (ρ_b and ρ_s), etc. A sensitivity analysis of this equation was undertaken with emphasis on soil salinity appraisal (Rhoades *et al.*, 1989c; Rhoades and Corwin, 1990). These findings are not reviewed herein. They show that the values of the soil parameters that can not be easily measured in the field (i.e., bulk density, particle density, clay percentage and total and "immobile" water contents) can be estimated sufficiently accurately for the purposes of practical soil salinity appraisal.

FIGURE 9
Correlation between EC_s and clay percentage for a number of soils from the San Joaquin Valley of California (after Rhoades *et al.*, 1989a)

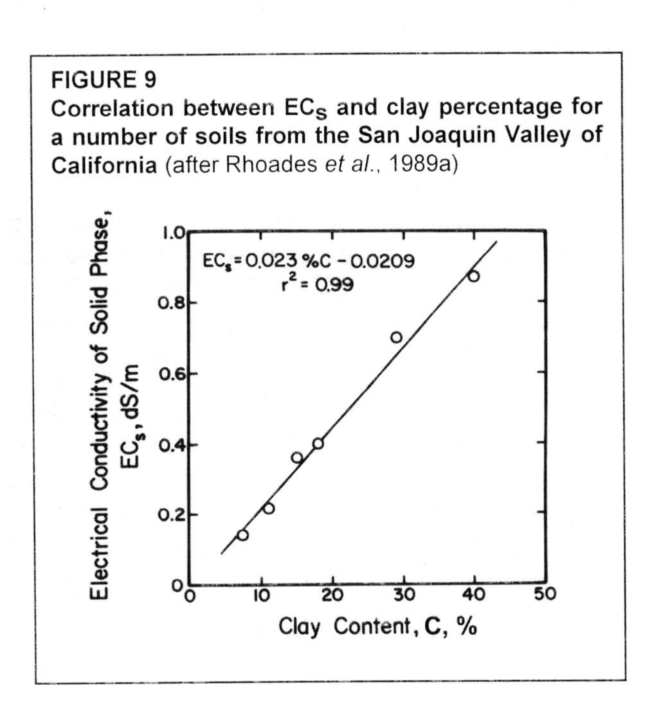

To use Equation [5] or [6] to assess soil salinity (EC_w or EC_e) from EC_a, the values of EC_s, Θ_{ws} and Θ_w must be known. EC_s, Θ_{ws} and Θ_{wc} can be estimated using Figures 9 and 10, respectively. The means used to obtain these relations are described elsewhere (Rhoades *et al.* 1989a). Θ_w can be measured in the field using various methods, if salinity is not too high, or it can be adequately estimated, for our purpose, by an indirect method that will be described later. Θ_s can be estimated from bulk density (ρ_B) as $\Theta_s \cong \rho_b /2.65$, where 2.65 is a reasonable estimate of the average particle density of most mineral soils. Bulk density can also be estimated sufficiently accurately for our purposes, as explained later.

FIGURE 10
Relationship between the volumetric content of soil water existing in the series-path (Θ_{ws}), the continuous-path (Θ_{wc}), and the total water content (Θ_w) found for various California soils (after Rhoades *et al.*, 1989a)

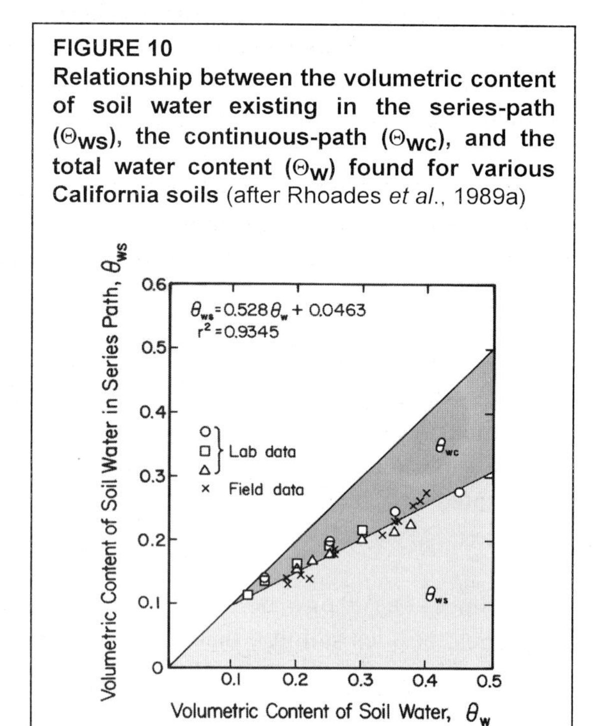

Another factor affecting EC_a, which may be important in some situations, is soil temperature. The electrical conductivity of soils containing moisture increases approximately 2 percent per degree Celsius increase in temperature. To simplify the interpretation of soil salinity data, it is customary to determine the temperature at which the measurement of EC_a is made, and then, by means of correction tables or equations, to convert the measurement to a reference temperature (Rhoades, 1976). The temperature factors (f_t) obtained from Equation [2] are suitable for this purpose (McKenzie *et al.*, 1989; Johnston, 1994, and Heimovaara (1995). However, sometimes it is preferable to encompass the effect of temperature by including it within the calibration relation established to predict soil salinity for the particular field conditions that existed at the time the EC_a measurements were made.

Equation [5] may be solved for EC_w, with the assumption that $EC_{ws} = EC_{wc}$, by arranging it in the form of a quadratic equation and solving for its positive root as:

$$EC_W = \frac{-b+\sqrt{b^2-4ac}}{2a}, \qquad [7]$$

where a = $[(\Theta_s)(\Theta_w - \Theta_{ws})]$, b = $[(\Theta_s + \Theta_{ws})^2(EC_s) + (\Theta_w - \Theta_{ws}) (\Theta_{ws} EC_s) - (\Theta_s EC_a)]$, and
$c = [\Theta_{ws} EC_s EC_a]$.

If EC_e is desired, it can be estimated from:

$$(EC_{wc} \Theta_{wc} + EC_{ws} \Theta_{ws}) = EC_w \Theta_w \cong EC_e \rho_b SP / 100, \qquad [8]$$

where SP is the gravimetric water content of the saturated-paste expressed as a percentage, and ρ_b is the bulk density of the soil (Rhoades, et al. 1989a). The latter derived relation is strictly valid only for chloride-salt systems. The errors inherent in this approximation are analogous to those discussed with reference to estimating EC_e from the EC value of extracts obtained at higher dilutions. However, the errors involved in Equation [8] are smaller because of the lower water contents used to make the saturation extract (compared to higher water/soil extracts). Some data, which supports the approximate equality of Equation [8], is presented in Figure 11. It may well be that EC_e values predicted from EC_a are more appropriate estimations of soil salinity than conventionally measured values of EC_e. This is so because the latter measurements are subject to the errors inherent to aqueous extracts previously discussed, because salts present within the "immobile" water contribute to EC_e but not to the EC_{wc} value, which the author considers the plant is more responsive to, and which mostly contributes to soil leachates.

The close relation that exists between EC_e and EC_a (as observed by numerous investigators) is made more apparent by substituting the above-mentioned approximate identity (Equation [8]) into equation [5], which yields the following relation:

$$EC_a = \left[\frac{(\Theta_s + \Theta_{ws})^2 \, EC_{ws} \, EC_s}{\Theta_s \, EC_{ws} + \Theta_{ws} \, EC_s} \right] + \left(\frac{\Theta_{wc}}{\Theta_w} \right) \left(\frac{SP \rho_b}{100} \right) EC_e . \qquad [9]$$

The experimental data presented in Figures 12 and 13 imply that this relation is generally applicable to arid-land soils. This relation also implies that the slope of $EC_a = f(EC_e)$ plots, and vice-versa, are related to soil type , since SP and ρ_b vary with soil type (evidence of this is shown in Figures 14 and 15). It also implies that the relation between EC_a and EC_e is not much influenced by variation in soil water content, as stated earlier, since the ratio Θ_{wc} / Θ_w is essentially a constant (~ 0.36; see Figure 10). The linear relations shown in Figures 13 and 14 were based on a relatively small data set of soils; they may be expected to be more curvilinear, like that shown in Figure 15, when a wider range of soil types are included. A curvilinear relation also has been reported by Johnston (1994).

FIGURE 11
Relationship between the product of soil water electrical conductivity (EC_w) and volumetric water content (Θ_w) and the product of the electrical conductivity of the saturated-paste extract (EC_e), its saturation percentage (SP), and the soil bulk density (ρ_b) found for soils of the Wellton-Mohawk Irrigation and Drainage District, Arizona, USA (after Rhoades, 1980)

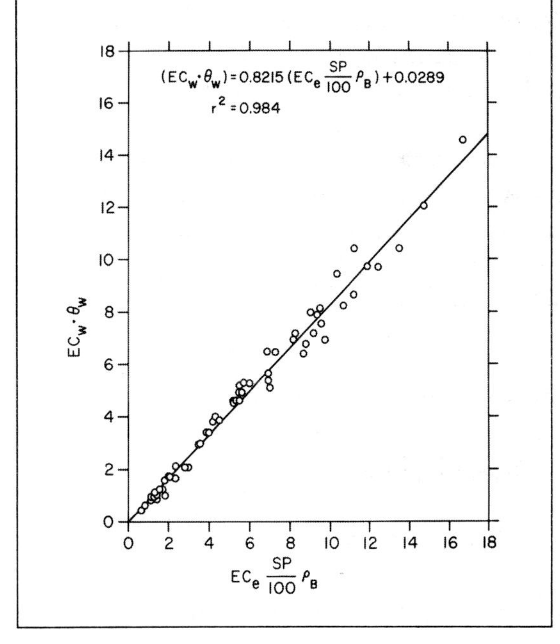

Equation [9] can be further simplified and approximated by substituting into it the value 0.36 for Θ_{wc} / Θ_w found to be typical of California arid-land soils and the familiar approximation (U. S. Salinity Handbook 60, 1954), $SP \rho_b / 100 = 2 \Theta_{fc}$, where Θ_{fc} is the volumetric water content at field capacity, to give:

$$EC_a = \left[\frac{(\Theta_s + \Theta_{ws})^2 \, EC_{ws} \, EC_s}{\Theta_s \, EC_{ws} + \Theta_{ws} \, EC_s} \right] + (0.72)(\Theta_{fc}) EC_e . \qquad [10]$$

Evidence to support this relation is given in Figure 16.

The above two relations, along with Equation [6], provide the theoretical basis for the previously reported findings that the calibration relating EC_a and EC_e for any saline soil can be expressed as a simple linear equation of the following type:

EC_e (or EC_w) = m (EC_a - EC_s^*)[11]

for which the slope can be predicted from their field capacity water content, or their SP value, or their texture using relationships like those shown in Figures 12 to 16 and the intercept value (EC_s^*, the intercept term of equations [5], [6], [9] and [10], essentially EC_s) can likewise be predicted from soil-texture related relationships such as that shown in Figure 9 (Rhoades, 1981; Rhoades, *et al.*, 1989a, 1990a). Of course, empirical calibrations of the type expressed by Equation [11] may be obtained for soils using various direct methods. Simple field-procedures have been developed in order to obtain calibration relations appropriate to field soils with their particular natural structures, pore size distributions and water holding properties (Rhoades, 1976, 1980, 1981; Rhoades and Ingvalson, 1971; Rhoades and van Schilfgaarde, 1976; Rhoades *et al.* 1977). Two of these procedures are illustrated in Figures 17 and 18 and are discussed in more detail in Annex 1.

FIGURE 12
Correlation between slopes of EC_e vs. EC_a calibrations obtained for different soils and their saturation percentage (SP) and bulk densities (ρ_b) (after Rhoades *et al.*, 1989a)

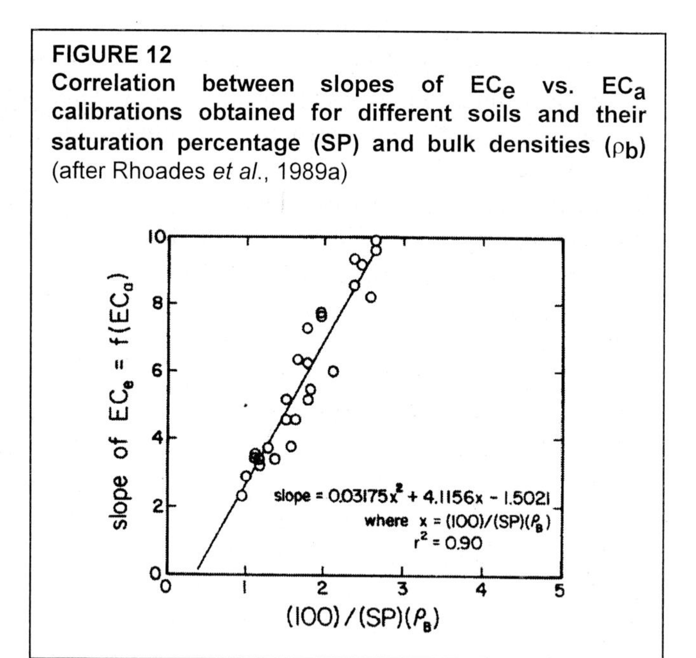

FIGURE 13
Correlation between slopes of EC_e vs. EC_a calibrations obtained for different soils and saturation percentages (after Rhoades, 1981)

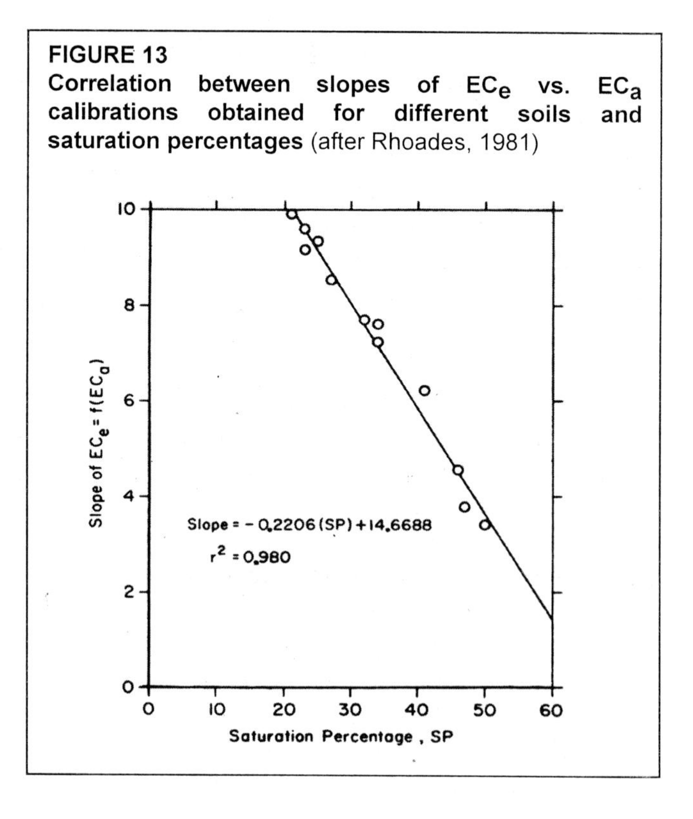

A typical linear relationship between EC_e and EC_a of the type expected from Equations [9] to [11], as obtained by the direct calibration of an arid-land soil at near field-capacity water content, is shown in Figure 19. Analogous relations between EC_w and EC_a have been developed for some typical arid-land soils (see Figure 20, after Rhoades, 1980). With such calibrations, one can predict EC_e (or EC_w) from EC_a for field soils of various types, provided they are in a sufficiently moist condition. While most of these calibrations have been developed for soils at or near field capacity water content at the time of EC_a measurement, they have also been developed for soils under drier conditions and found not to differ substantially (Halvorson and Rhoades, 1974).

Numerous satisfactory field calibrations (like Figure 19) have been obtained for many soils around the world and they have been found to be similar for soils of similar textures (Rhoades and Ingvalson, 1971; Halvorson and Rhoades, 1974; Rhoades, 1976, 1979a, 1980, 1981; Halvorson *et al.* 1977; Rhoades *et al.* 1977, 1989a; Yadav *et al.* 1979; Loveday, 1980; van Hoorn, 1980; Nadler, 1981; Bohn *et al.* 1982; Johnston, 1994). These calibrations have been found to be essentially independent of soil sodicity, provided soil structure and porosity have not been seriously degraded by the sodicity. Evidence of this is given in Figure 21 obtained in the controlled laboratory experiments of Bottraud and Rhoades (1985a). The sodium adsorption ratio (SAR) is a good estimator of soil sodicity (US Salinity Laboratory, 1954). Additional supportive evidence, though less rigorous and exact, is found in Shainberg *et al.* (1980) and Johnston (1994).

An alternative model procedure for determining salinity from EC_a at various water contents has been suggested by Nadler (1982). This procedure "curve-fits" what amounts to a f = (Θ) relation using moisture-tension data established for the particular soil in question and an empirical "effective porosity" relation based on ΔΘ. To date, the method has been successfully applied only to disturbed soil samples; it requires considerable laboratory effort to establish the empirical fit; it only applies to the "fitted" soil, and its applicability to field soils was found to be not generally good (unpublished data).

FIGURE 14
Relationship found between saturation percentage and clay percentage for some California soils (after Rhoades *et al.*, 1990a)

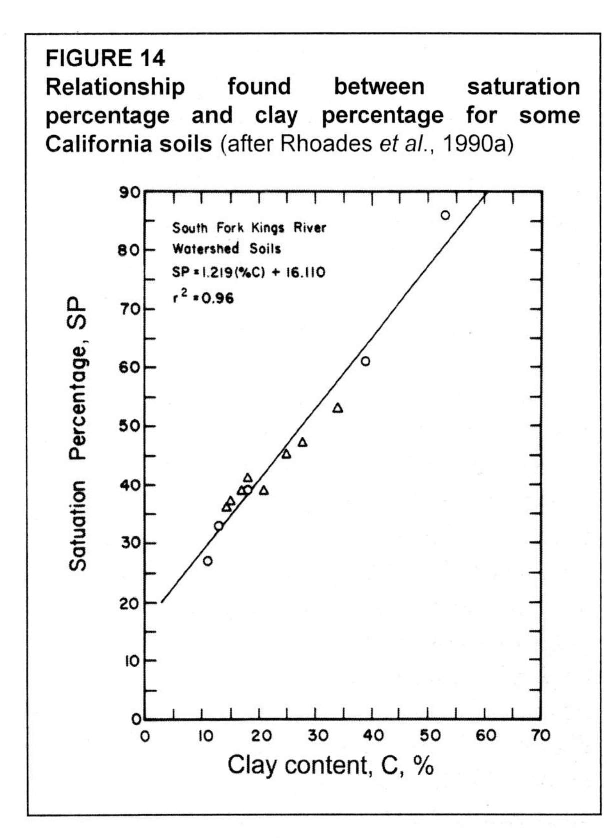

FIGURE 15
Relationship between slopes of EC_e vs. EC_a calibrations obtained for different soils of Montana, USA, and their clay contents (after Halvorson *et al.*, 1977)

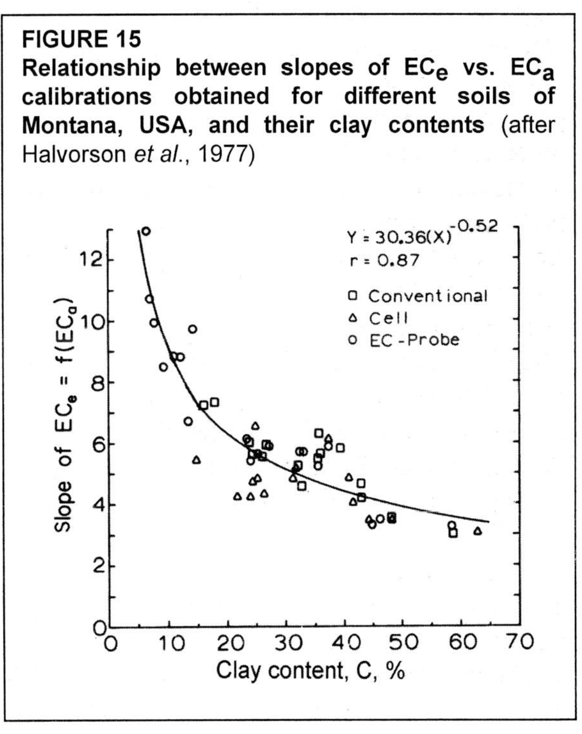

The advocated model and experimental data (Rhoades *et al.*, 1976, 1989a, 1989c; Rhoades, 1990b) show that EC_a is primarily a measure of **the content of dissolved electrolyte** present in a unit-volume of soil; note that the product (EC_{wc} * Θ_{wc}) is analogous to the product of concentration of soil water times volume of mobile soil water. Salt-free water is not a significant conductor of electricity; hence, the water in the soil is simply the "container" of the mobile electrolyte (the dissolved salt) and the "conduit" for the flow of electricity. Therefore, the effect that changes in water content have on EC_a measurements and salinity appraisal depends on whether or not salt loss occurs with the change in water content. Immediately following an irrigation or rain event, salt removal from the soil occurs as the water content drains to "field capacity"; hence, measurements of EC_a are relatively sensitive to changes in Θ_w during such times because the product ($EC_w * \Theta_w$) decreases in proportion to the change in water content. However, the soil is usually too wet during this period to permit one to access the field and to undertake measurements of EC_a. These measurements generally only become feasible later and are typically made after the rapid drainage has ceased and the soil is at, or below, field capacity water content. During this latter period, further major losses of soil water in the rootzones of cropped soils occur mainly through transpiration and almost all of the salt contained in the water retained by the soil following irrigation is left behind in the remaining soil water; hence, the salt concentration (and EC_w) of the remaining water is increased approximately proportionately to the reduction in Θ_w and, thus, the product ($EC_w * \Theta_w$) is approximately constant (hence, EC_a) is fairly independent of changes in water content **following drainage**. Thus, in saline soils, EC_a is much more related to salinity (as expressed in terms of EC_e or EC_w) than to water content and is, **for any given irrigated and cropped soil**, not much affected by the changes in water content that occur in the time period between the cessation of rapid drainage and the next irrigation event, i.e., during the period of time when measurements of EC_a are normally made. The relatively small changes in EC_a that do occur with changes in Θ_w under these conditions result from a variation in the partitioning of soil water between Θ_{ws} and Θ_{wc} (previously referred to in Rhoades *et al.*1976, as a change in tortuosity) and from the precipitation of some salt as the solubilities of calcite and gypsum are exceeded. As Θ_w decreases below field capacity due to evapotranspiration, EC_a will show a relatively small and approximately linear decrease according to the relationship:

FIGURE 16
Correlation between slopes of EC_e vs. EC_a calibrations obtained for different soils and their field-capacity water contents (after Rhoades, 1981)

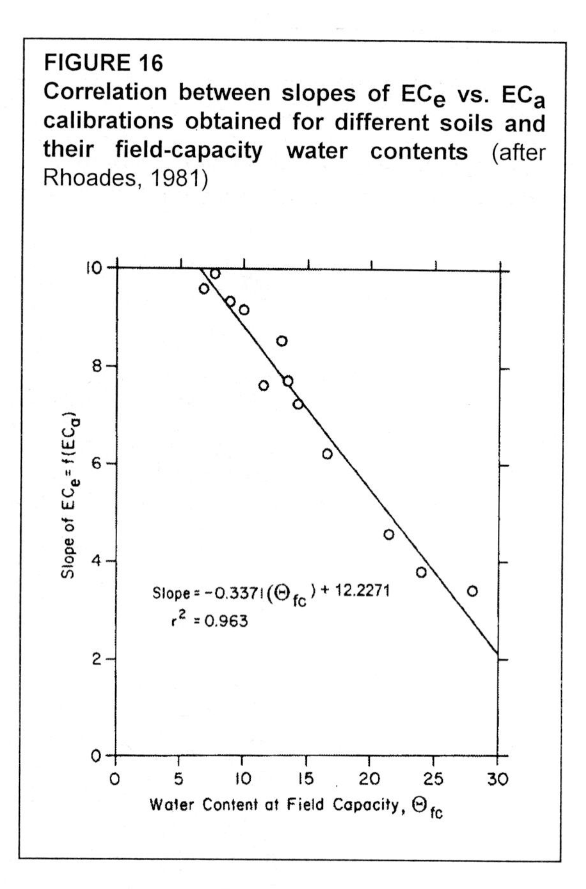

$$\Delta EC_a = \alpha \, \Delta\Theta_w \; K, \qquad [12]$$

where α is a factor related to the relation between Θ_{wc} and Θ_{ws}, and $K = EC_w \, \Theta_w \cong$ a constant. For typical soils the error in EC_a caused by $\alpha \, \Delta\Theta_w$ is not large with reasonable deviation in Θ_w from field capacity water content. Experimental evidence to support the above "argument" made on the basis of theory and logic have been obtained in both laboratory and field studies; some are reported in Rhoades *et al.*, 1981, 1989a, 1989c, 1990a; Rhoades and

FIGURE 17
(A) Cylinder and surrounding "moat" with impounded saline water used to leach the soil and adjust it to a desired level of salinity; (B) access-hole being made in soil with Oakfield-type soil sampling tube for subsequent insertion of EC_a-probe; (C) EC_a-probe being inserted into salinity-adjusted soil for determination of EC_a; and (D) sample of salinized soil being collected for subsequent determination of EC_e (salinity) (after Rhoades *et al.*, 1977)

Corwin, 1990; Rhoades, 1990b; Bottraud and Rhoades, 1985b). Example data supporting this conclusion are given in Figure 22, after Rhoades *et al.* (1981). Indirect evidence supporting this conclusion are the low correlations typically found between EC_a and Θ_w in any given irrigated, salt-affected field but for which high EC_a - EC_e correlations are found. Johnston (1994) concluded in a test of this conclusion that the "compensation" described above was valid, but less than what the author found. However, he did not carry out his study in cropped fields where the "drying" mechanism is water removal by roots that are distributed throughout the entire relatively large volume of soil. He subjected small soil columns to the drying action of air which would be expected to cause water (and salt) to flow to the ends of the column away from where the electrodes were located. This mass flow is analogous to that which causes the changes in water and salt contents that occur in soils during drainage that, as discussed earlier, result in changes in EC_a because the **product** of (EC_{wc} * Θ_{wc}) is reduced; whereas, changes in water content caused by transpiration do not reduce this product except when salt solubilities are exceeded.

The appropriateness of using Θ_w as a reference for water content and of the inappropriateness of using matric potential (such as tensiometer readings) in establishing EC_e = f (EC_a) calibrations is supported by the results of the laboratory column-studies of Bottraud and Rhoades (1985b). These data are not

FIGURE 18
Soil-filled, four-electrode cell (as obtained with a coring device) showing one group of four of the eight electrodes inserted into the undisturbed soil used to measure EC_a; after the soil is removed, it is analysed in the laboratory for EC_e. (after Rhoades *et al.*, 1977)

FIGURE 19
Relationship between bulk soil electrical conductivity and electrical conductivity of the saturated-paste extract for Dateland soil at field capacity water content (after Rhoades, 1980)

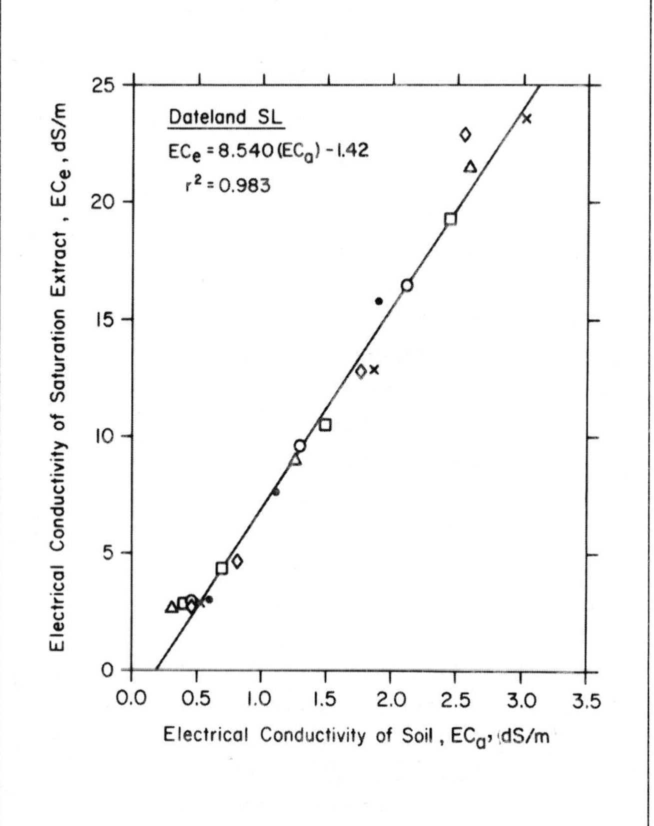

reviewed here. Suffice it to say that EC_a is directly related to Θ_w in the manner previously described (and not to matric potential by any general relation) and if matric potential were used as a reference one would also need to know whether the soil was in a wetting or drying cycle.

As shown above and in the previous section, EC_a is highly influenced by salinity and, for any given soil, is not much influenced by normal variations in water content encountered during practical measurement times. However, as shown in equations [9] and [10], **for a given salinity**, EC_a increases as Θ_{fc}, SP (which itself increases with clay and organic matter contents) and ρ_b increase, because of their effects on the slope term, and as EC_s, Θ_s, and Θ_{ws} increase, because of their effects on the intercept term. EC_s will increase as the clay content, cation exchange capacity and organic matter content of the soil increase. Θ_{ws} will increase with increases in clay content, organic matter content and bulk density, which itself generally decreases with increases in clay and organic matter. Thus, it is evident that soil texture and organic matter content, and correlated soil properties, will influence EC_a and, in the absence of salinity, can be expected to be capable of being determined from sensor measurements of EC_a, or EC_a*, so long as the soils contain enough water to provide a continuous pathway for electrical current flow. The use of the mobilized sensor-surveys / "stochastic-calibration" approach described later is a very practical, efficient and accurate methodology for establishing such EC_a - soil property correlations and for developing

FIGURE 20
Relationships between bulk soil electrical conductivity and soil water electrical conductivity for the major soils of the Wellton-Mohawk Irrigation Project of Arizona USA (after Rhoades, 1980)

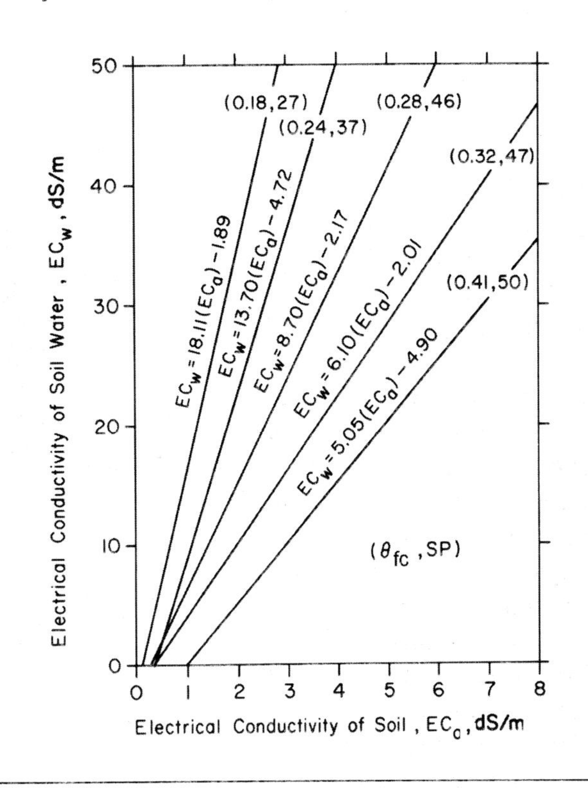

FIGURE 21
Relationship between the soil electrical conductivity of Fallbrook soil and the electrical conductivity and sodium adsorption ratio (SAR) of the soil water (after Bottraud and Rhoades, 1985)

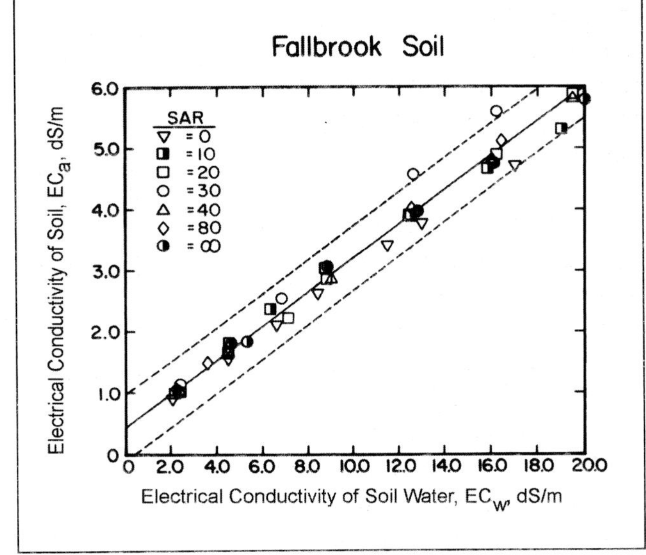

FIGURE 22
Response of soil electrical conductivity to irrigations and evapotranspiration over several irrigation cycles (after Rhoades *et al.*, 1981)

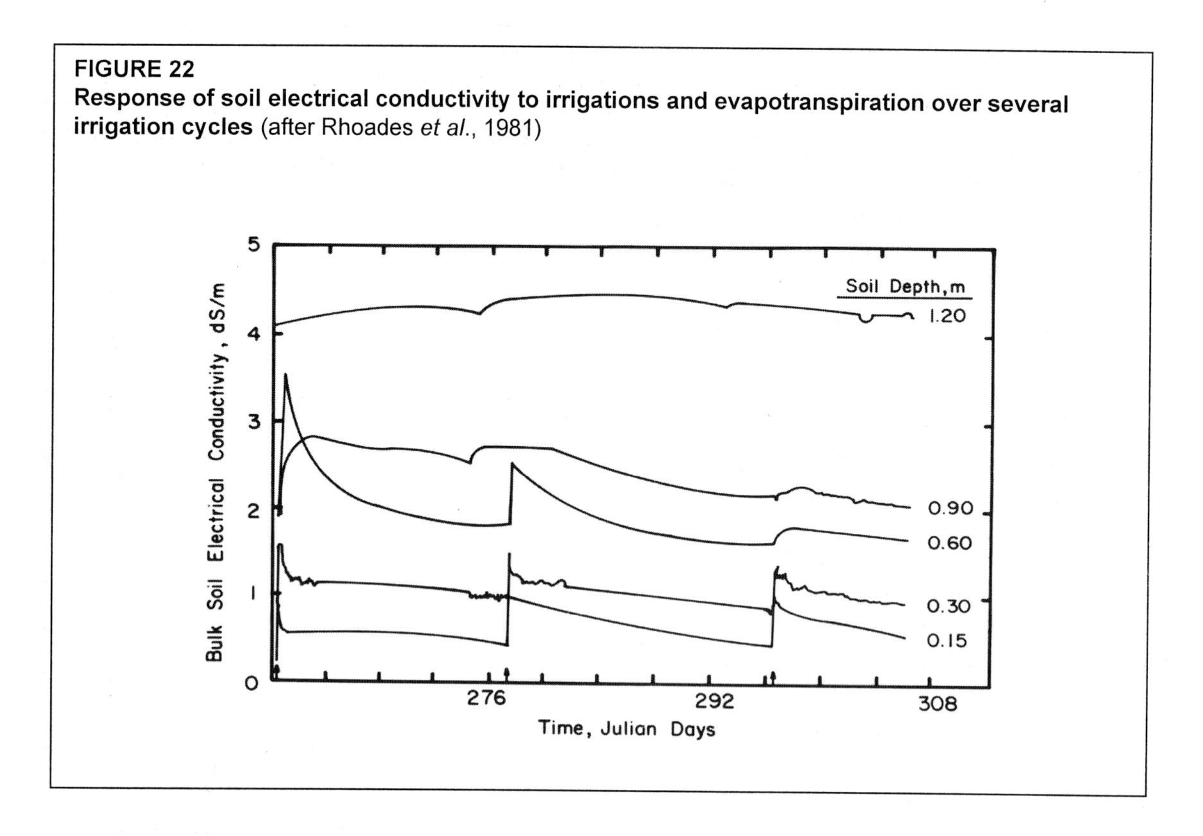

much of the soil-property information required for prescription farming purposes (Rhoades *et al.*, 1997d). This prescription farming application is discussed more later.

An equation analogous to that of [5] established for bulk soil electrical conductivity has been developed (Rhoades *et al.* 1989b) for saturated soil-pastes, as follows:

$$EC_p = \left[\frac{(\Theta_s + \Theta_{ws})^2 \; EC_{ws} \; EC_s}{(\Theta_s) \; EC_{ws} + (\Theta_{ws}) \; EC_s} \right] + (\Theta_w - \Theta_{ws}) \; EC_e \, , \qquad [13]$$

where the EC of the equilibrated extracted solution (EC_e) is analogous to EC_{wc} , EC_p is the electrical conductivity of the saturated-paste, Θ_w and Θ_s are the volume fractions of total water and solids in the paste, respectively, Θ_{ws} is the volume fraction of water in the paste that is coupled with the solid phase to provide a series-coupled electrical pathway through the paste, EC_s is the average specific electrical conductivity of the solid particles, and the difference (Θ_w - Θ_{ws}) is Θ_{wc}, which is the volume fraction of water in the paste that provides a continuous pathway for electrical current flow through the paste (a parallel pathway to Θ_{ws}). Assuming the average particle density (ρ_s) of mineral soils to be 2.65 g/cm^3 and the density of saturated soil-paste extracts (ρ_w) to be 1.00, Θ_w and Θ_s for saturated pastes can be directly determined from SP as follows:

$$\theta_w = SP / \left[\left(\rho_w 100 / \rho_s \right) + SP \right], \qquad [14]$$

and

$$\Theta_s = 1 - \Theta_w \, . \qquad [15]$$

The saturation percentage of most mineral soils can be adequately estimated in the field, for purposes of salinity appraisal, from the weight of a known volume of paste (Rhoades *et al.* 1989b). Figure 23 may be used for this purpose; for details of the relationships inherent in this figure see Wilcox (1951). Evidence of the validity of this is shown in Figure 24.

These relationships can be used to determine soil salinity using soil samples. The method requires the creation of saturated soil-pastes but avoids the need for the collection of the extract. Calibrations are needed for each different soil, but they are easily and accurately predicted by the means described in the next section. The method is faster and more field-practical than the conventional extraction procedures.

DETERMINING SOIL SALINITY FROM SATURATED SOIL-PASTE ELECTRICAL CONDUCTIVITY

EC_e can be determined from measurements of EC_p and SP (using equations [13] to [15]), if values of ρ_s, Θ_{ws} and EC_s are known. These parameters can be adequately and simply estimated, as demonstrated by Rhoades *et al.* (1989b & c). For typical arid land soils of the Southwestern United States, ρ_s may be assumed to be 2.65 g/cm^3; EC_s may be estimated from SP as: $EC_s = 0.019\ (SP) - 0.434$ (see Figure 25), and the difference (Θ_w - Θ_{ws}) may be estimated from SP as: $(\Theta_w - \Theta_{ws}) = 0.0237\ (SP)^{0.6657}$ (see Figure 26). The measurement of EC_p and SP (from the volume-weight of the paste and Figure 23) can be easily made using an EC-cup of known geometry and volume, conductance meter and battery operated balance as shown in Figure 27. This permits EC_e to be

FIGURE 23
Theoretical relation between saturation percentage (SP) and weight (in grams) of 50 cm^3 of saturated paste, assuming a particle density of 2.65 g/cm^3 (after Wilcox, 1951 and Rhoades *et al.*, 1989b)

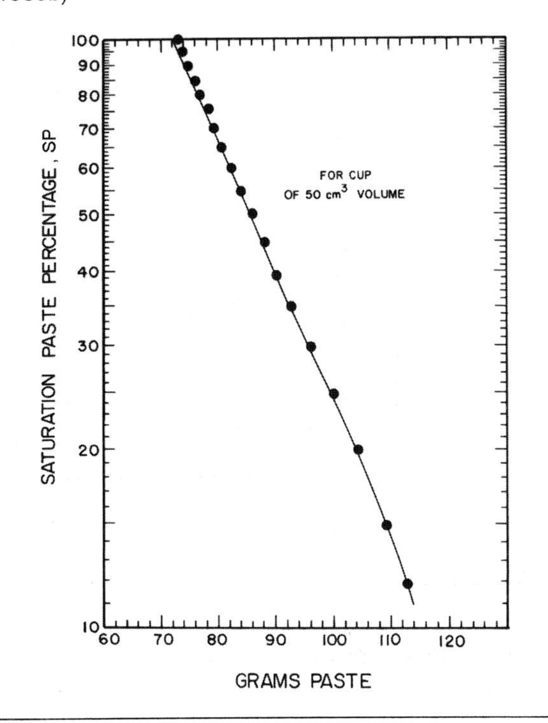

FIGURE 24
Correspondence between measured and estimated (using Figure 23) saturation percentages, for a set of California soils (after Rhoades *et al.*, 1989b)

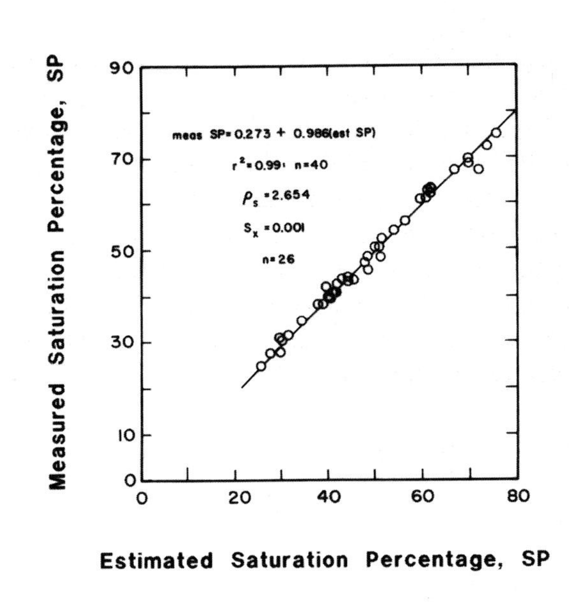

determined from equations [13] to [15] (as described below, or from Figure 28) from the simply made measurements of the volume-weight and the EC of the saturated soil-paste. Evidence of the validity of this is shown in Figures 29 and 30. The method is suitable for both laboratory and field applications, especially the latter, because the apparatus is inexpensive, simple and rugged and because the determination of EC_p can be made much more quickly than with the conventional procedure which involves the vacuum-extraction of the paste and, subsequently, the measurement of the EC of the extract. An additional savings in time occurs because less soil needs to be sampled (**100 grams** is sufficient, which can be collected relatively quickly with a Lord-type, or similar, sampling **tube**) and less saturated-paste needs to be prepared (it takes about 1/3 the time to prepare a saturated-paste for this method compared to the conventional method which requires about **400 grams** and collection with a more time-consuming soil **auger**).

FIGURE 25
Correlation between the electrical conductivity of the soil solid phase (EC_s) and the saturation percentage (SP) for some soils of the San Joaquin Valley of California USA (after Rhoades *et al.*, 1989a)

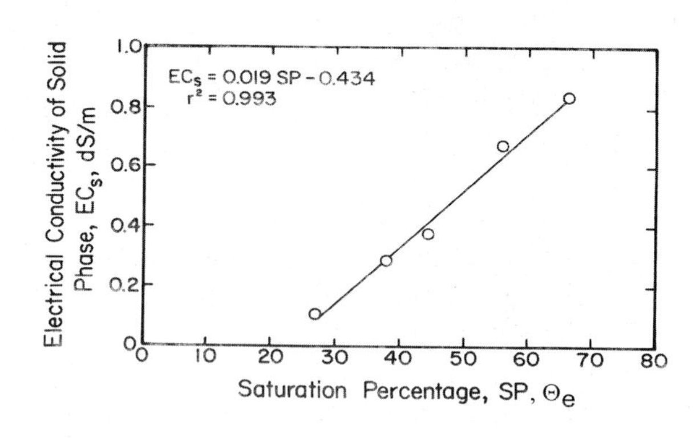

FIGURE 26
Relationship between the volumetric content of water in the saturated soil-paste which is in the continuous electrical conduction path (Θ_w - Θ_{ws}) and the saturation percentage (SP) for some San Joaquin Valley California USA soils (after Rhoades *et al.*, 1989b)

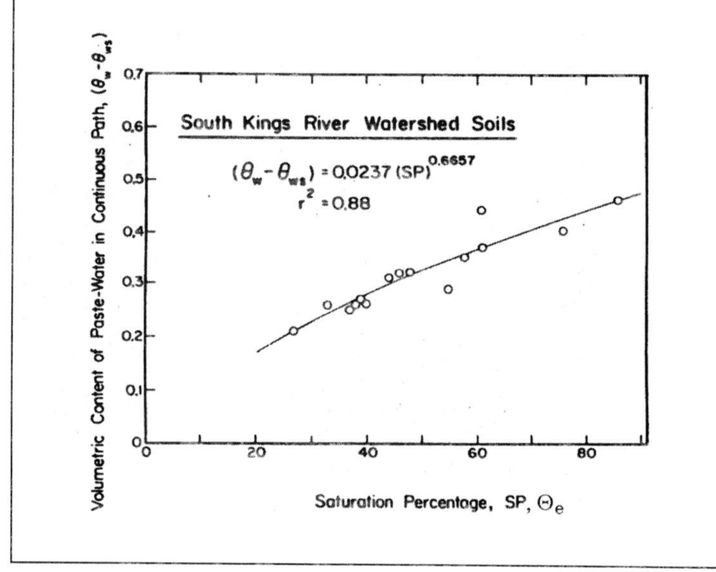

As indicated above, EC_e may be estimated from Figure 28 given EC_p and SP, using the curve corresponding to the appropriate SP value, or else it may be calculated using the following equation:

$$EC_e = \frac{-b + \sqrt{b^2 - 4ac}}{2a}, \qquad [16]$$

FIGURE 27
(A) Portable balance used in the field to determine the weight of the saturated soil-paste filling the "Bureau of Soils Cup", (B) "Bureau of Soils Cup" filled with saturated soil-paste connected to conductance meter, and (C) close up of "Bureau of Soils Cup" (after Rhoades 1992)

where $a = [\Theta_s (\Theta_w - \Theta_{ws})]$, $b = [(\Theta_s + \Theta_{ws})^2 (EC_s) + (\Theta_w - \Theta_{ws}) (\Theta_{ws} EC_s) - (\Theta_s) EC_p]$, and $c = -[(\Theta_{ws}) (EC_s) (EC_p)]$. The values of EC_s, Θ_s, Θ_w and Θ_{ws} are estimated from SP using the relationships described above.

Sensitivity analyses and tests have shown that the estimates used in this method are generally adequate for purposes of salinity appraisal of typical mineral arid-land soils (Rhoades *et al.* 1989c). This method has been found to be quite accurate and robust (considerable experience and data have been obtained using it in the salinity assessment research of the US Salinity Laboratory). As discussed earlier, this method is a better choice to estimate EC_e than those calculated from 1:5, and similar, soil:water extracts. For organic soils, or soils of very different mineralogy or magnetic properties, these estimates may be inappropriate. For such soils, appropriate values for Θ_s, EC_s and Θ_{ws} will need to be determined using analogous techniques to those of Rhoades *et al.* (1989b). The accuracy requirements of these estimates may be evaluated using the approaches given in Rhoades *et al.* (1989c).

It should be noted that ($EC_e\ \Theta_e$) is not equivalent to ($EC_w\ \Theta_w$) because different amounts of soil are involved in the two measurements. The relationship between these two products is:

$$EC_w\ \Theta_w / \rho_b = EC_e\ \Theta_e / \rho_p. \qquad [17]$$

Data to support this is given in Rhoades (1981) and Rhoades *et al.* (1990). The ratio Θ_e / ρ_p is equivalent to SP/100 (see Rhoades *et al.* 1989a, b).

The procedure described in this section is especially suitable to determine the EC_e values of the soil samples used to calibrate the stochastic-model (described later) which, in turn, is used to predict soil salinity from sensor measurements of EC_a.

This paste-method of determining EC_e has been commercialized in the United States (see Table 12, Chapter 5) and is available in a kit form, including software for the calculations. This company also sells an analogous field-kit to determine soil sodicity (in terms of SAR_e) without the need for collecting extracts or performing analyses of calcium and magnesium concentrations, based on the methodology of Rhoades *et al.*,

FIGURE 28
Relationships between EC_p, EC_e and SP for representative arid-land soils (after Rhoades *et al.*, 1989b)

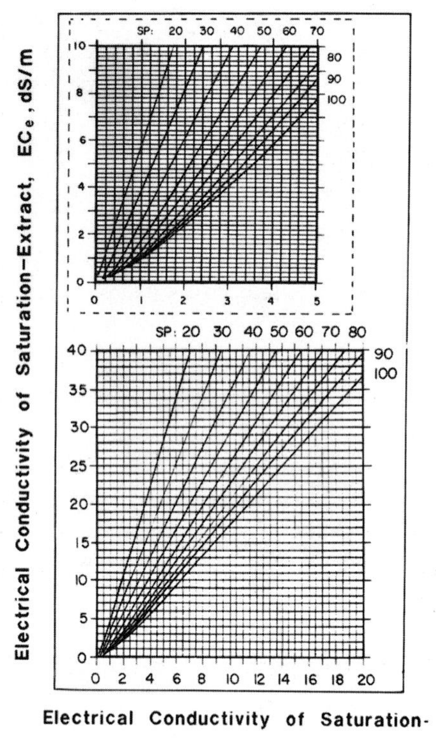

FIGURE 29
Relationship between EC_p and EC_e for Grangeville soil. The symbols represent empirical data and the solid line is the "fit" of these data using Eq. [13] (after Rhoades *et al.*, 1989b)

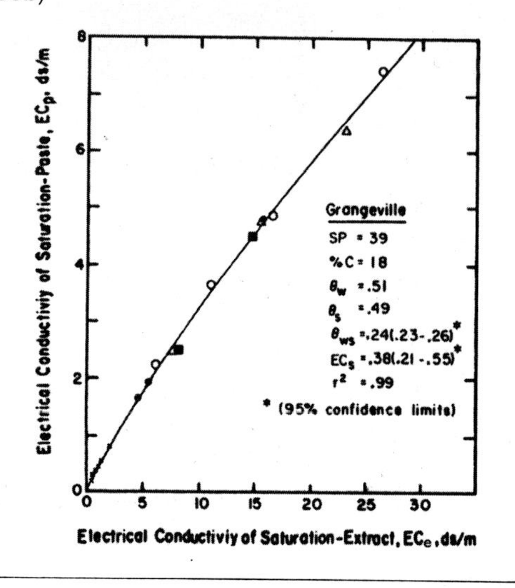

(1997c). The sodicity method, while not as accurate as the salinity method, is sufficiently accurate for many field diagnosis purposes and, most certainly, for screening samples to identify those that merit the time, labour and expense of laboratory analyses. These two kits permit soil salinity and sodicity to be determined directly in the field using saturated-pastes of soil samples.

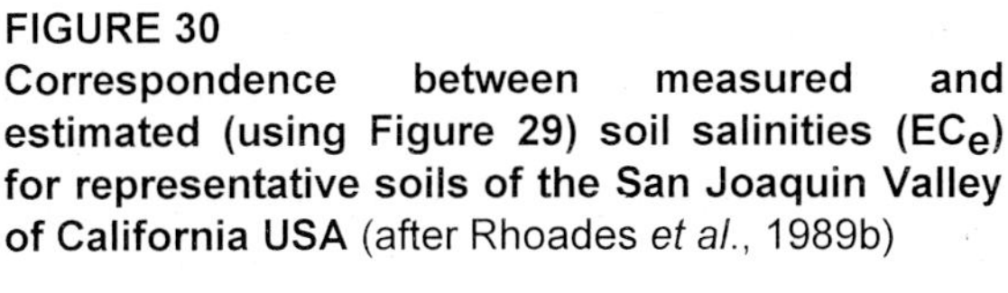

FIGURE 30
Correspondence between measured and estimated (using Figure 29) soil salinities (EC_e) for representative soils of the San Joaquin Valley of California USA (after Rhoades *et al.*, 1989b)

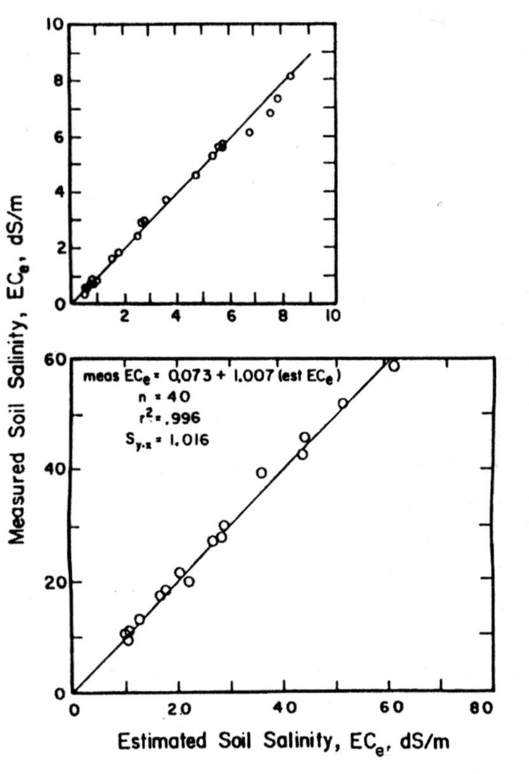

DETERMINING SOIL SALINITY FROM BULK SOIL ELECTRICAL CONDUCTIVITY

Soil salinity can be determined from measurements of bulk soil electrical conductivity using essentially three different approaches. After reviewing the various instrumental means of measuring EC_a, this section discusses these alternative methods of salinity appraisal.

Sensors and equipment for measuring soil electrical conductivity

Three types of sensors are commercially available for measuring bulk soil electrical conductivity. Two are field-proven, portable sensors: (i) four-electrode sensors and (ii) electro-magnetic induction sensors. A third sensor-type, based on time-domain-reflectometry (TDR) technology, has not yet been shown to be sufficiently accurate, simple, robust or fast enough for the general needs of field salinity assessment. Each of the first two sensor-types has its' own advantages and limitations. The first two sensors and related equipment will now be described and discussed.

Four-electrode Sensors

Bulk soil electrical conductivity can be measured using four-electrodes inserted into the soil, a combination electric-current generator/resistance, or conductance meter, and connecting wire. A photograph of the basic "surface-array" equipment is provided in Figure 31. With such equipment the depth and volume of measurement may be varied by changing the spacing between the current (outside) electrodes, as illustrated in Figure 32. When the distance between the outside pair of electrodes (the current electrodes) is small, the flow of electricity is shallower than when the distance is greater. The effective depth of measurement is about one-third of the distance between current electrodes (see Figure 33). The spacing-depth relation is discussed in more detail later. The calculation of EC_a from surface-array measurements requires knowledge of the spacing between the current and potential (inner pair) electrodes. An equation for calculating the "cell constants" for different arrays and spacings of electrodes is given later.

The current source-meter unit may be either a hand-cranked (see Figure 31) or a battery-powered type (see Figure 34). Some of the available generator/meters were designed for geophysical purposes and these generally read in ohms of resistance. If such units are to be used for purposes of general soil salinity assessment, they should measure from 0.1 to 1000 ohms. A unit specifically developed for soil salinity appraisal is available from Martek Instruments[1]. It is battery powered, allows the geometry constant to be set for different configurations of electrodes and reads out directly in terms of soil electrical conductivity (an early model is shown in Figure 34; it is also more portable (smaller/lighter) than the typical units built for geophysical prospecting purposes). One can build a simple, but adequate, generator/meter unit from components using the circuitry-schematic and parts-list provided in Austin and Rhoades (1979) and in Annex 2. Units which can data-log time and EC_a measurements have also been developed for use with both hand-held four-electrode sensors and with a mobilized, tractor-mounted version of a "fixed-array" unit developed for making automated "on-the-go" measurements of bulk soil electrical conductivity. The latter unit and mobilized system are described below.

FIGURE 31
Four electrodes positioned in a Wenner-type surface-array and a combination electrical generator and resistance meter (after Rhoades and Oster, 1986)

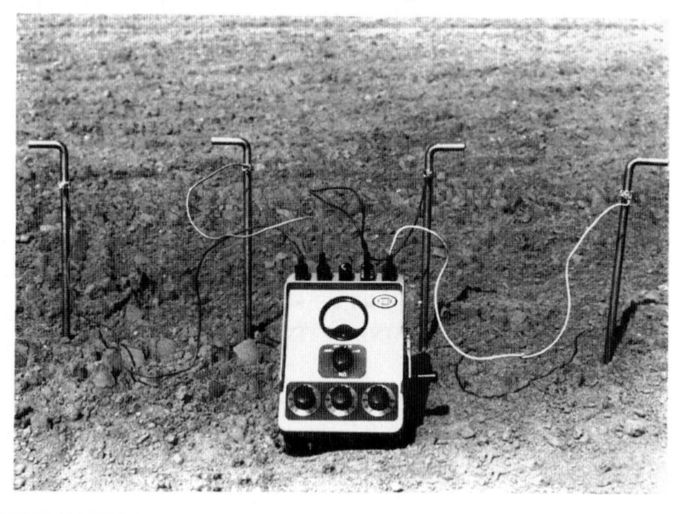

FIGURE 32
Schematic showing increased depth and volume of EC_a measurement with increased C_1-C_2 electrode spacing. Effective depth of measurement is approximately equal to one-third of (C_1 - C_2). C stands for current-electrode and P stands for potential-measuring electrode (after Rhoades, 1976)

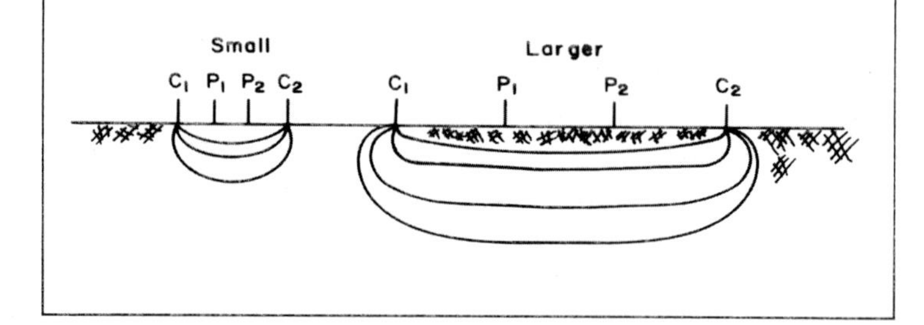

Electrodes used in surface arrays of the type shown in Figures 31 and 34 can be made of stainless steel, copper, brass, or almost any other corrosion-resistant, conductive metal. Array electrode size is not critical, except that the electrode must be small enough to support itself when inserted to a depth of 5 cm or less. Electrodes 1.0 to 1.25 cm in diameter by 45 cm long are convenient for most measurement purposes, although smaller electrodes are preferred for determining EC_a within soil depths of less than 30 cm. The effect of depth of insertion of the electrodes is discussed later; an equation useful in this regard is given in Annex 3. Any flexible, well-insulated, multi-stranded, 12 to 18 gauge wire is suitable for connecting the array-electrodes to the meter.

For hand-carried mapping or traverse work, it is convenient to mount the array-electrodes in a board with a handle (see Figure 34) so that soil resistance, or conductivity, measurements can be made relatively quickly. Such mounted-units are practical for current-electrode spacings of up to about two metres; switching devices have been developed to make it easy to switch the meter quickly between the different sets of electrodes (Rhoades, 1976). These "fixed-array" units save the time involved in spacing the electrodes and keep the "geometry factor" constant from one measurement site to another.

A mobilized, tractor-mounted version of a "fixed-array" four-electrode unit has been developed for making automated "on-the-go" measurements of bulk soil electrical conductivity. Generator/meter/logger units which can data-log time and EC_a measurements have also been developed for use with this mobilized equipment, as well as with hand-held four-electrode sensors. The mobilized four-electrode system also data-logs the associated locations in the field, as determined using global positioning system (GPS) equipment. This system, shown in Figure 35, is capable of making both faster and wider-spaced readings than can be accomplished manually, while simultaneously providing the x, y coordinates of each measurement site. It is especially well suited for collecting detailed information about the variability of average root zone soil electrical conductivity within fields and of the various soil properties that can be inferred from EC_a.

FIGURE 33
Variation of: (A) current density with depth in a plane mid-way between the current electrodes; (B) current density at unit depth as a function of current electrode separation (after Rhoades and Ingvalson, 1971)

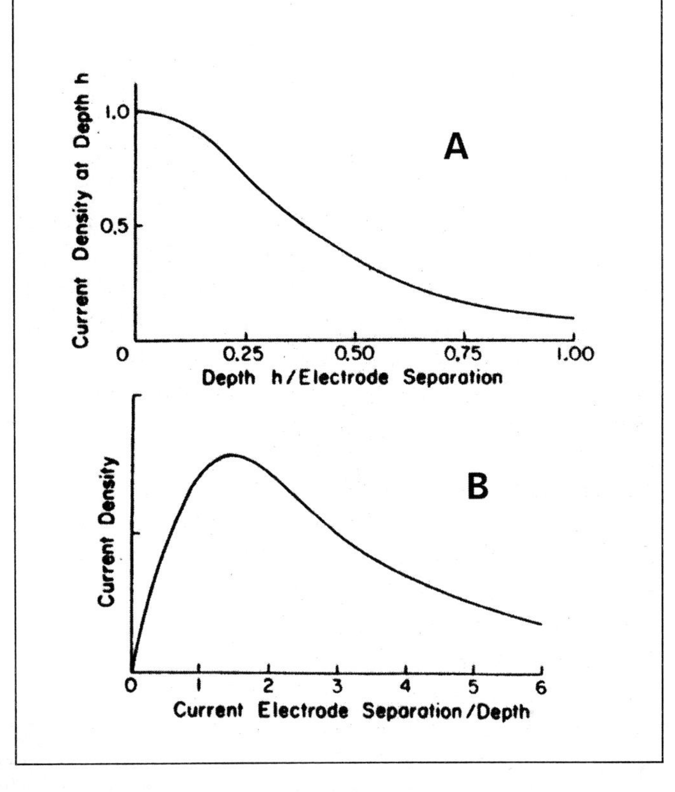

FIGURE 34
A"fixed-array" four-electrode apparatus and commercial generator/meter (after Rhoades, 1978)

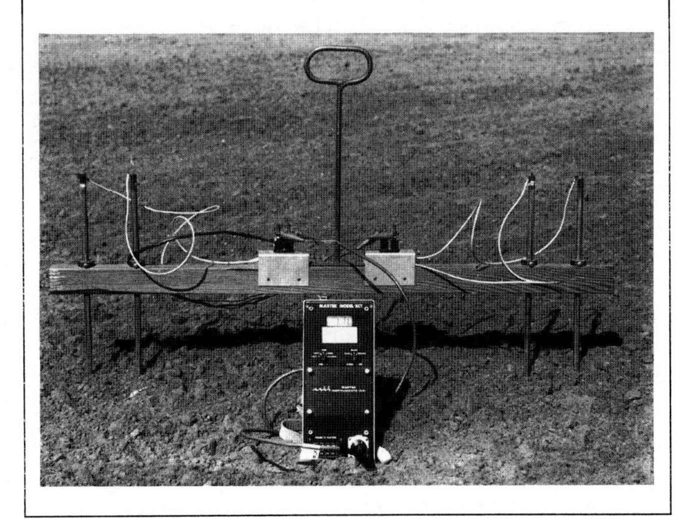

A close up view of one of the electrodes used in this system is shown in Figure 36. The use of the latter equipment to assess soil salinity and related management is described in more detail later and elsewhere (Rhoades, 1992a, 1992b, 1993, 1994, 1996b; Carter *et al.*, 1993; Rhoades *et al.*, 1997a, 1997b). The electrodes and generator/meter can be attached to the tool bar of almost any tractor, or they can be provided "together with a dedicated tool bar and weather proof container for the generator/meter system (as shown in Figure 35). This mobilized/ automated four-electrode system is commercially available from Agricultural Industrial Manufacturing, Inc.[1]. A more recently commercial mobilized four-electrode unit, though less well suited for use in furrow-irrigated cropped fields, is available from Veris Technologies[1].

FIGURE 35
A mobilized (tractor-mounted) "fixed-array" four-electrode system, with mast for GPS antenna (after Rhoades, 1992)

FIGURE 36
(A) Close-up of the four fixed-array electrodes used on the mobilized tractor-mounted system, (B) insulators used to isolate sensor part of shank from the rest of the tractor, and (C) close up of replaceable pad at bottom of electrode (after Carter *et al.*, 1993)

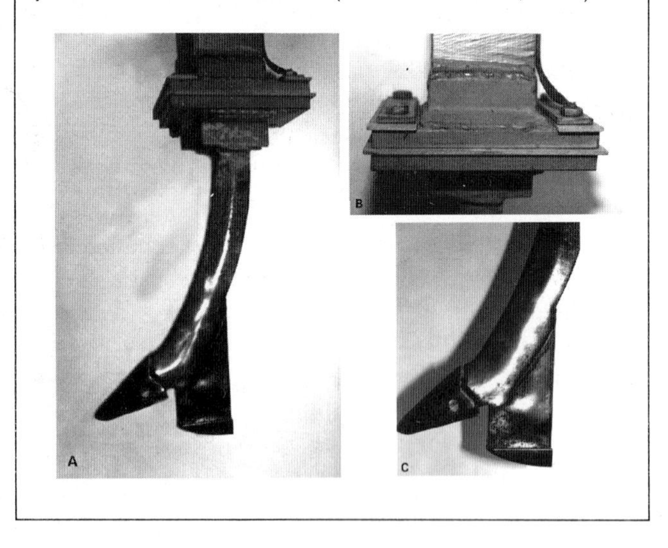

A salinity probe, in which the four electrodes are incorporated into a shaft, was developed by Rhoades and van Schilfgaarde (1976). With this probe, EC_a can be measured in small soil-volumes and at various depths in the soil profile. Convenient sized current source-meter units have been designed for use with the four-electrode salinity (Austin and Rhoades, 1979). Commercial versions of both the four-electrode probe and "meter" are made by Martek Instruments[1], by Eijkelkamp Agrisearch Equipment[1] and by Elico Limited[1]. The probe and meter sold by "Eijkelkamp" are essentially the same as those developed by Rhoades and collaborators. The units sold by Martek Instruments are improved versions. The newest version of the Martek SCT meter, which reads directly in EC_a corrected to 25 °C and which incorporates a data logger and a timer is shown in Figure 37, along with the moulded, insertion salinity-probes (both standard- and micro-sizes) they sell. Versions of the four-electrode probes could be, but have not yet

[1] Mention of trademark or proprietary products in this manuscript does not constitute a guarantee or warranty of the product by the Food and Agriculture Organization of the United Nations and does not imply its approval to the exclusion of other products that may also be suitable.

been, developed that are suitable for incorporation into a mobilized, automated system; however they can presently be used to advantage, even if manually, for some detailed assessment purposes (such as characterizing the salinity patterns within seedbeds and through root zones). Burial-type four-electrode units (see Figure 38) suitable for monitoring applications are also available from Martek Instruments. Simpler units can be built as originally developed by Rhoades (1979). Information in this regard is given in the Annex 4. Other special-purpose cells have been built for measuring EC_a of undisturbed soil cores or in laboratory soil columns/cells. These units are shown in Annex 5.

FIGURE 37
Two commercial four-electrode probes (small and standard sizes) and electrical generator-meter/data-logger (after Rhoades, 1992).

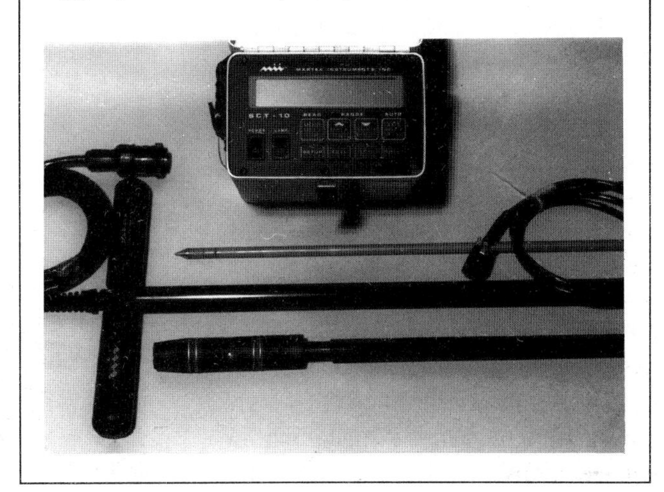

FIGURE 38
Commercial burial-type four-electrode conductivity probe used for monitoring changes in soil electrical conductivity (after Rhoades and Corwin, 1984)

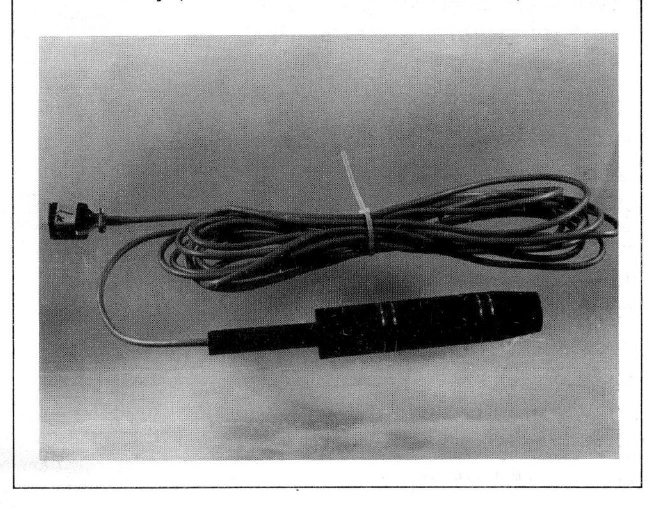

Electromagnetic-induction Sensors

Soil electrical conductivity can be measured remotely using electromagnetic induction (EM) methodology. The basic principle of operation of the EM soil electrical conductivity meter is shown schematically in Figure 39. An EM transmitter coil located in one end of the instrument induces circular eddy-current loops in the soil. The magnitude of these loops is directly proportional to the electrical conductivity of the soil in the vicinity of that loop. Each current loop generates a secondary electromagnetic field that is proportional to the value of the current flowing within the loop. A fraction of the secondary induced electromagnetic field from each loop is intercepted by the receiver coil of the instrument and the sum of these signals is amplified and formed into an output voltage which is linearly related to **depth-weighted** soil electrical conductivity, EC_a^*. The nature of the depth weighting is discussed later.

Figure 40 shows a commercially available EM soil salinity sensor (Geonics EM-38[1]) oriented in both the horizontal (EM_H; Figure 40A) and vertical (EM_V; Figure 40B) coil-positions. This device was designed, at the request of the author, to meet the general-purpose needs of soil salinity appraisal (McNeill, 1992). The EM-38 device contains appropriate circuitry to minimize instrument response to the magnetic susceptibility of the soil and to maximize response to electrical conductivity. It has an inter-coil spacing of 1 metre, operates at a frequency of 13.2 kHz, is powered by a 9 volt battery, and reads out directly in terms of EC_a^*.

The coil configuration, frequency and inter-coil spacing were chosen to permit measurement of EC_a^* to effective depths of approximately 1 and 2 metres when placed at ground level in horizontal and vertical configura-tions, respectively. Other "EM" units are available which are capable of deeper measurement. For more details about the principles of EM measurements and the various sensors that can be used in this regard, see McNeill (1980 and 1992).

Mobilized, automated EM-measurements can be made within various depths of the rootzone using the EM-38 sensor, as well as with a combined four-electrode sensor if desired, and the mobilizing/ auto-mating equipment developed by Rhoades and collaborators shown in Figure 41 (Rhoades, 1992a, 1992b, 1993, 1994, 1996b, Carter *et al.*, 1993). With this system, some 52 operator-actions are automatically performed to collect a sequence of EM-38, four-electrode, and GPS readings at a given site. These actions are made in less than one minute (about 20 sites per hour, including travel time, can be sampled with this system). With these data, the salinity level and distribution within the soil profile to be determined in two dimensions. The EM-38 is contained within the cylinder protruding in front of the mobilizing unit. The sensor is in the "up" (travelling) position in this picture. The optional four-electrode component is shown in Figure 42 in the "down", inserted position. As with the mobile, four-electrode system previously described, this system also incorporates synchro-nized, GPS site-positioning equip-ment and data logging capabilities. This mobilized/automated system is

FIGURE 39
Schematic showing the principle of operation of an electromagnetic induction soil conductivity sensor (after McNeill, 1980)

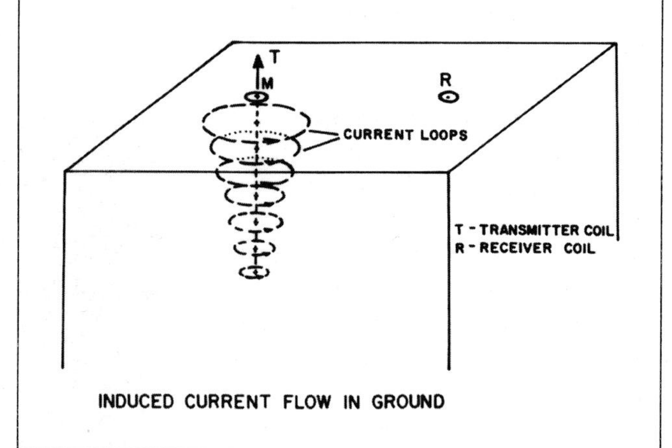

FIGURE 40
Geonics EM-38 electromagnetic soil conductivity sensor in (A) horizontal orientation and (B) vertical orientation (after Rhoades, 1992)

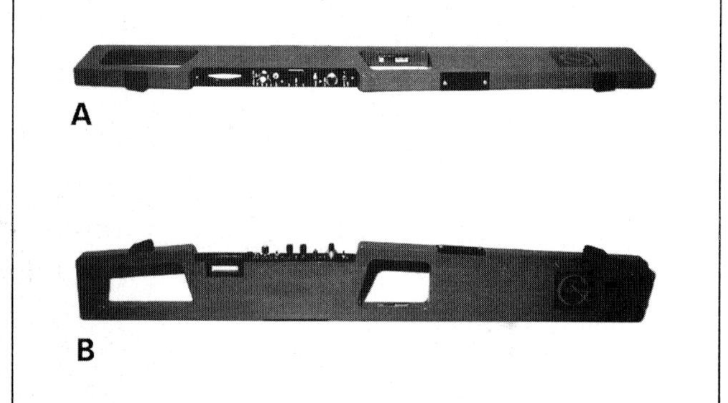

FIGURE 41
Mobilized salinity assessment system with combined EM-38 and four-electrode soil conductivity sensors and mast for mounting a GPS antenna. Both sensors are in the 'up' travel position (after Rhoades, 1992; 1993; Carter *et al.*, 1993)

very well suited for the detailed mapping of EC_a* and correlated soil properties, as well as for the mapping of these properties within different depth-intervals of the root zone and slightly deeper. The system is commercially available from AIM Inc.[1]. It also incorporates a laser mast and a load-cell which permits the four-electrode array to also function as a penetrometer (not shown). Thus, additional, complementary information can be simultaneously obtained about the micro-relief features of the field and about the compaction and crusting properties of the soil surface. The latter co-located information helps determine the cause(s) of salinization, and the appropriateness of irrigation/drainage management, considering the observed salinity levels and patterns. Several prototypes of this system have been designed and tested; they are described in more detail in Rhoades (1992a, 1992b), Rhoades, (1993), Carter *et al.* (1993), Rhoades (1994), Rhoades (1997b), and Rhoades *et al.* (1997a, 1997b, 1997d). Other forms of mobilization of EM-sensors have been undertaken, though they are not as integrated nor as well adapted to row-cropped fields as the above described system (Cameron *et al.*, 1994; Jaynes, 1996; *Kitchen et al.*, 1996).

FIGURE 42
Close-up of the fixed-array four-electrode unit in (A) the travel-position and (B) inserted into the soil by the hydraulic "scissors" apparatus of the mobilized combination sensor assessment system (after Carter *et al.*, 1993)

Procedures for Measuring Bulk Soil Electrical Conductivity

Large-volume Measurements

For the purpose of determining soil salinity within root zones, or some fraction thereof, it is desirable to make the measurement of EC_a to depths of up to 1 to 1.5 metres. This may be accomplished with both four-electrode and EM-sensors. It is accomplished with the four-electrode equipment by configuring the surface-array of electrodes in a straight line with the spacing between the two outer (current) electrodes selected depending upon the desired depth(s). As implied in Figure 32, the depth and volume of measurements are readily altered by varying the spacing between the current-electrodes. The relative spacing between the inner (potential)-electrode pairs can also be varied, but this does not affect the depth of measurement. The electrodes are often spaced in the so-called Wenner-array with equal spacings between all of them (Wenner, 1916; Rhoades and Ingvalson, 1971). When using the Martek[1] SCT meter, each of the inner-pair of electrodes is preferably placed inward from its closest outer-pair counterpart a distance equal to 10 % of the spacing between the outer-pair. In both of the above-mentioned electrode-arrangements, as well as for others, the inner-pair of electrodes is generally used to measure the electrical potential (or resistance) while current is passed between the outer-pair. The effective depth of current penetration for either configuration (in the absence of appreciable soil layering) is approximately equal to about one-third the outer-electrode spacing, y; thus "average" soil salinity is measured to a depth equal to approximately y/3 (Rhoades and Ingvalson, 1971; Rhoades, 1976; Halvorson and Rhoades, 1976). Thus, by varying the spacing between current electrodes, one can measure salinity to different depths, also of different volumes, in soil using the four-electrode system.

An advantage of this "surface-array" method is the relatively large volume of soil that is measured compared to that of the insertion four-electrode probes (discussed later) or of customary soil samples. The volume of measurement is about $(y/3)^3$, where y is as defined above. Hence, effects of small-scale variations in field-soil salinity can be minimized by these relatively large-volume measurements.

For measurements taken in the Wenner-array (electrodes equally spaced) using geophysical type meters, which measure resistance, bulk soil electrical conductivity is calculated, in dS/m, as:

$$EC_a = 159.2 f_t / a R_t, \qquad [18]$$

where a is the distance between the electrodes in cm, R_t is the measured resistance in ohms at the field temperature t, f_t is the previously described temperature compensating factor used to adjust the reading of EC_a to a reference temperature of 25° C, and 159.2 is the numerical equivalent of 1000/2 π. For measurements made with other spacings of electrodes, EC_a is calculated (after Dobrin, 1960) as:

$$EC_a = \left(1000 / 2\pi R\right) \left\{ f_t / \left[\frac{1}{\frac{1}{r_1} - \frac{1}{r_2} - \frac{1}{R_1} + \frac{1}{R_2}} \right] \right\}, \qquad [19]$$

where R is resistance, f_t is the temperature correction factor, and r_1, r_2, R_1 and R_2 are the distances in cm between various pairs of electrodes (Figure 43). For measurements made with the Martek SCT meter, the meter is calibrated for any set of electrode spacings by setting it to read the value of EC_a calculated from Equation [19] while the corresponding resistance is connected between the outer-electrodes. Use of a variable resistor box is convenient in this regard.

FIGURE 43
Schematic of distances between the current and potential electrodes in four-electrode array for use with Equation [19] (after Dobrin, 1960)

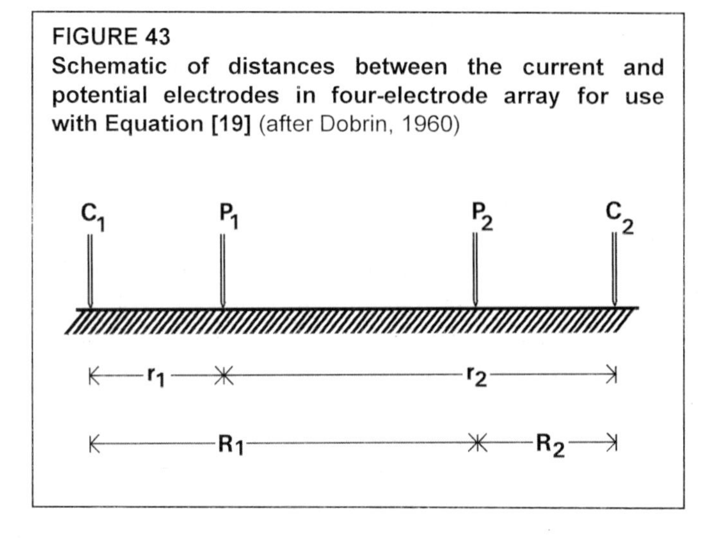

In both Equation [18] and [19], the electrodes are assumed to make only point contacts with the soil. In practice they must be inserted into the soil far enough to support their weight and to make contact with soil having sufficient moisture to permit a meaningful measure of EC_a and interpretation of salinity. An equation is provided in Annex 3 to correct for this depth of insertion, assuming the soil is uniform throughout this depth and the rest of the soil volume involved in the measurement. Since this assumption is seldom true, the "correction" equation given in Annex 3 is seldom used in practice. Instead, an attempt is made to minimize the depth of insertion, especially for shallow soil depth measurements. Based on empirical findings of Rhoades and Ingvalson (1971), the depth of insertion should be no more than 25 mm for measurements within the 0-0.3 m soil depth and no more than 50 mm for measurements within the 0-0.6 m soil depth. For deeper soil depth measurements, the electrodes may be inserted up to depths of 75 mm with no discernible effect. The diameter and length of the electrodes should be reduced when measurements are to be made with them inserted to shallow depths; otherwise, the larger electrodes can not be supported by the soil and make good contact. They tend to fall over and to "break contact" with the soil.

In most field situations, the immediate topsoil is too dry and loose to attempt to make measurements when the electrodes are inserted to shallow depths. Hence, they are inserted to a depth below the boundary separating the dry/loose surface mulch and the underlying moist/firm soil, at whatever depth this boundary occurs. This depth is then taken as the "zero-depth" for inferring soil salinity from the EC_a measurements, or for collecting soil samples to be used to establish EC_e - EC_a calibrations for the soil, since no useful relation exists between salinity and EC_a in dry soils as explained earlier. The fact that four-electrode measurements do not apply to the dry/loose surface mulch through which the electrodes were inserted in order to reach moist/firm soil may be seen as either a serious limitation of the surface-array, four-electrode method, compared to the EM method/sensor (Johnston, 1994), or as an advantage for reasons given later. Unfortunately, the incorporation of such surface soil, especially when it is highly salinized through evaporation-driven processes, in soil samples collected to calibrate or test four-electrode systems (also EM systems) has resulted in some erroneous calibrations, misinterpretations, and conclusions reported by a number of users and even investigators. It needs to be stressed that the variability that exists over very short distances in surface irrigated soils is often very great. Hence, great care must be made to avoid the collection of samples from regions of the soil that are not within the volume of the four-electrode sensor measurement when establishing calibrations or applying/testing them. A good example of such short-scale variability is shown in Figure 73, which is presented and discussed later. It needs to be recognized, as mentioned earlier, that what is many times used as "truth" regarding soil salinity in testing the models, data, various sensors, approaches and methodology involved in salinity assessment is, in fact, often not an adequate/appropriate index of salinity; certainly not that involving dry soil samples and high water:soil ratio extracts. Special care must be taken to account for the spatial variability that exists in typical saline soils. The mobilized measurement systems and stochastic-model approach of salinity assessment, which is discussed later, were explicitly developed to provide practical tools to measure and characterize spatially variable soil salinity in irrigated fields. Examples will be given later to illustrate the utility of these systems and approach to describe and account for this spatial-variability dilemma.

Relatively large volumes of soil can also be measured with the EM-38 sensor, though less than with the tractor-mounted, fixed-array, four-electrode system. The volume and depth of EM-sensor measurements are influenced by the spacing between coils, the current frequency, and the orientation of the axes of the magnets/coils with respect to the soil surface plane (McNeill, 1980). The effective depths of measurement of the Geonics EM-38[2] device are about

1 and 2 metres when it is placed on, or in close proximity to, the ground and the coils are positioned horizontally and vertically, respectively. The effective width of the measurement extends out about ½ metre to the sides and ends of the unit (McNeill, 1990). Thus the elliptical volume of the measurement has an length of about 2 m, a width of about 1 m and a depth that corresponds to the "Z" values given in Figure 44 and described by the equations provided in footnote number 2. The depth (and associated volume) does not contribute equally to the measurement, as explained next.

FIGURE 44
Cumulative relative contribution of all soil electrical conductivity, R(Z), below various depths sensed by the EM-38 unit when placed on the soil surface in horizontal (parallel) and vertical (perpendicular) magnetic-coil positions
(after McNeill, 1980)

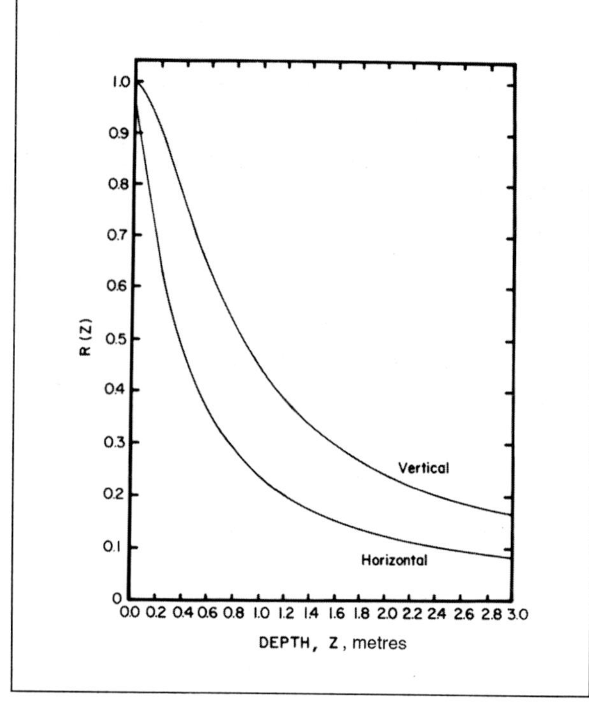

The EM-38 device does not provide a linear measure of EC_a with depth, rather a depth-weighted value EC_a^* is obtained as stated earlier. The theoretical depth-distributions of this weighting for an **homogeneous** soil are shown in Figure 44 for both the vertical (EM_V) and horizontal (EM_H) configurations. The ratio of these distributions is shown in Figure 45. These distributions show that the EM_H and EM_V measurements are not independent; they are interrelated measurements, though not so much in the shallow depths (such as 0-0.30 m). The 0 to 0.3, 0.3 to 0.6, 0.6 to 0.9, and 0.9 to 1.2 m soil depth intervals contribute about 43, 21, 10, and 6 percent, respectively, to the EC_a^* reading of the EM unit when it is positioned on homogeneous ground in the horizontal position (Rhoades and Corwin, 1981). Thus, the depth-weighted bulk soil electrical conductivity read by the EM device for this situation and in this configuration is approximately:

$$EC_a^* = 0.43EC_{a,\,0\text{-}0.3} + 0.21EC_{a,\,0.3\text{-}0.6} + 0.1EC_{a,\,0.6\text{-}0.9} + 0.06EC_{a,\,0.9\text{-}1.2} + 0.2EC_{a,\,>1.2}, \qquad [20]$$

where the subscript designates the depth interval in metres. Corresponding percentages in the vertical position are 17, 21, 14 and 10, respectively[2]. Recent studies show that these proportions do not hold for non-homogeneous profiles (Rhoades *et al.* 1990b). The EM_H and EM_V measurements made with the EM-38 measurements also depart differently from the actual EC_a values, even in homogeneous soils, when these levels exceed about 2 dS/m, as shown in Figure 46. For these reasons, the "profiling" methods based on theoretical, uniform-profile weighting functions (such as that of Slavich, 1990) are not generally reliable in their applications. Methods that account for the effect of non-uniform weighting functions need to be used. Some of these methods are discussed/referenced later.

[2] The relative contributions (R) to the secondary EM field (or EC_a^*) from all material below a depth can be theoretically calculated from $R_V = 1/(4^2 + 1)^{1/2}$, and $R_H = (4^2 + 1)^{1/2} - 2$, for the vertical (V) and horizontal (H) dipoles, respectively (McNeill, 1980).

The EM-38 does not lend itself well to the direct determination of average EC_a (or EC_e) since its response is weighted by depth (generally decreasing with depth). However, as pointed out by Rhoades and Corwin (1981), the depth-weighted reading it provides (Equation 20]) is close to that which many researchers regard as the way crops extract soil water within their root zone and the way they proportionately respond to the variation of salinity with depth in the root zone. For this reason, Rhoades and Corwin (1981) suggested that the EC_a^* reading obtained with the EM-38 might provide a reasonable measure of crop-effective salinity. Based on this concept/suggestion, Wollenhaupt *et al.* (1986) developed depth-weighted calibrations for the EM-38 and soils they worked with in Canada using salinity values by depth in the soils weighted in accordance with slightly modified versions of the depth-response relations given above. Subsequently, McKenzie *et al.* (1989) developed analogous but slightly different weighted-calibration relations for the EM-38 for use with their soils, as did Johnston (1994) in South Africa. There is some evidence that crop response to salinity can be reasonably related to the depth-weighted reading provided by the EM-38 (Slavich and Read, 1983; McKenzie *et al.*, 1990; Rhoades *et al.*, 1997d). In spite of this, the latter three sets of calibrations referenced are subject to the same criticism given above about the inapplicability of the theoretical depth-response of the EM sensors (which are based on uniform-depth salinity conditions) to depth-varying conditions of soil salinity. However, they can be used advantageously for applications where **relative** differences in salinity need to be mapped in landscapes; but one should not expect to accurately determine the salinity distribution through the soil profile using such calibrations.

FIGURE 45

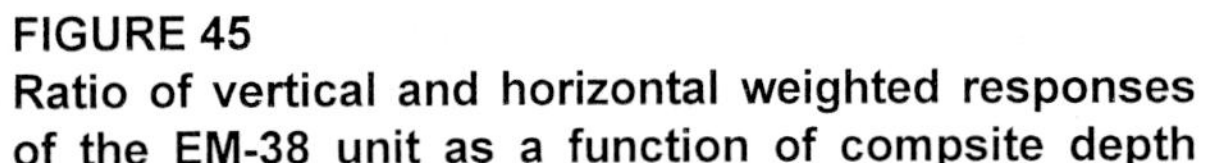

Ratio of vertical and horizontal weighted responses of the EM-38 unit as a function of compsite depth increments (i.e. 0-0.15, 0-0.30, 0-0.45, 0-0.60 m, etc.) (after Corwin and Rhoades, 1990)

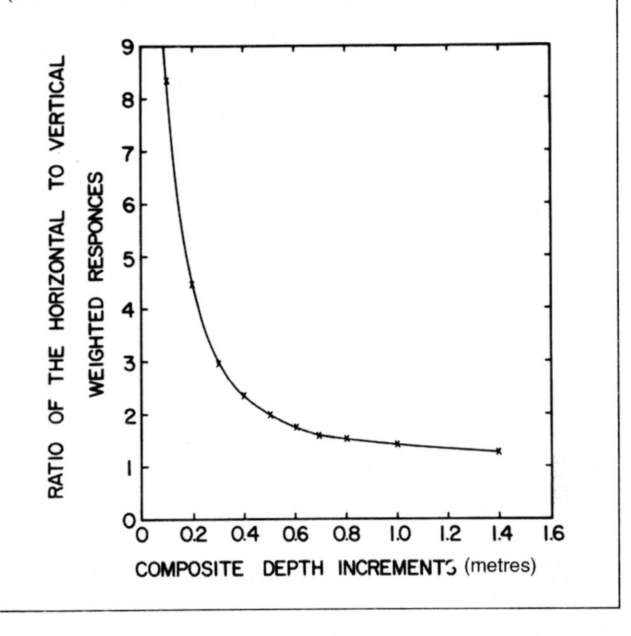

FIGURE 46

EM-38 readings for homogeneous profiles as a function of the profile EC_a value (personal communication from J. D. McNeill; Rhoades, *et al.*, 1990)

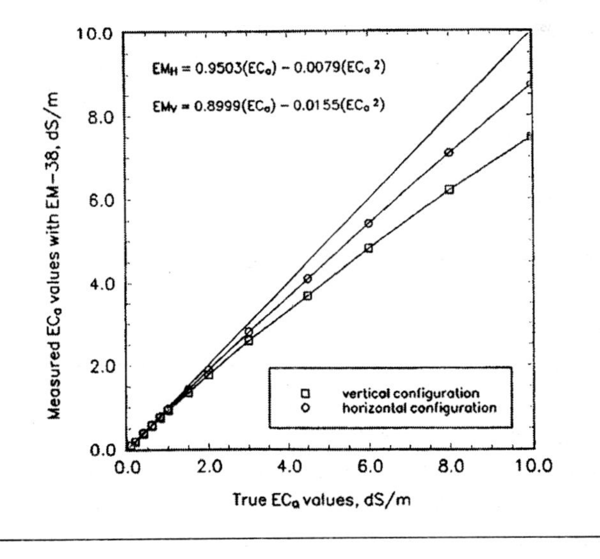

It is often desirable to be able to determine soil EC_a within various depth intervals so that soil salinity levels within the various parts of the root zone can be calculated, as needed for making certain assessments and management decisions. Since the proportional contribution of each soil depth interval to the EM-38 reading (EC_a^*) can be varied by changing the coil orientation or, as shown by Rhoades and Corwin (1981), by raising the unit to various heights above the ground, it is possible to estimate EC_a within various depth-increments of the soil from a succession of EM measurements made at various orientations, or at various heights above-ground, or both. The EC_a values within different discrete soil depth intervals of a group of California soils have been found to be correlated with a succession of EM_H readings made above ground as:

$$EC_{a,\ 0\text{-}0.3} = \alpha_0 EM_0 + \alpha_1 EM_1 + \alpha_2 EM_2 + \alpha_3 EM_3 + \alpha_4 EM_4, \quad [21a]$$

$$EC_{a,\ 0.3\text{-}0.6} = \beta_0 EM_0 + \beta_1 EM_1 + \beta_2 EM_2 + \beta_3 EM_3 + \beta_4 EM_4, \quad [21b]$$

where EM represents the reading obtained with the EM-38 unit held in the horizontal position and 0, 1, 2, 3 and 4 represent height above ground in increments of 30-cm. The author has found these relations to apply in each of the several widely distant places in the world where he has personally tested them. But since they have not been widely tested and may be expected to vary for different EC_a - depth patterns, for the reasons described above, they should be used with caution until they have been evaluated for the specific conditions of interest.

Subsequently, another series of empirical equations and coefficients were developed to estimate EC_a within discrete soil depth intervals using just two measurements made with the magnetic coils of the EM-38 instrument positioned at ground level, first horizontally and then vertically (Corwin and Rhoades, 1982, 1984). These equations were initially developed using the same relatively small data set as that involved in the development of Equation [21]. Salinity increased with depth in all of these soil profiles. Subsequent tests showed the new empirical relations were inaccurate when applied to soils with salinities which decreased with depth; hence, a new set of coefficients were developed for such soils (Corwin and Rhoades, 1984). The two categories of soils were distinguished by the EM_V / EM_H ratio. Profiles in which salinity increased with depth were called "regular" profiles and were associated with EM_V / EM_H ratios of ≥ 1. Profiles in which EC_a (salinity) decreased with depth were called "inverted" profiles and were associated with EM_V / EM_H ratios of < 1. The derivation of these relations and the resulting general form of the equation are given in Annex 6. These relations were subsequently modified/improved using a substantially larger data base (Rhoades *et al.* 1989d) and expressed in the following form:

$$(EC_{a,\ x1\text{-}x2})^{0.25} = k_H (EM_H)^{0.25} + k_V (EM_V)^{0.25} + k_3, \quad [22]$$

where x1 - x2 represents a given depth increment in cm, EM_V and EM_H are the readings obtained with the EM-38 device positioned at the soil surface in the vertical and horizontal positions, respectively, k_H, k_V and k_3 are empirically determined coefficients for each depth increment, and the exponent 0.25 is an empirical factor used to provide a more normally-distributed set of values. This approach based on but two EM-38 readings is more practical to use than the approach inherent in Equation [21] which requires five measurements. Equation [22] is also more easily solved than is Equation [21] and the results were found to be almost as accurate for the two depth intervals 0 - 30 and 30 - 60 cm, when tested using the same original data set. Johnston (1994) evaluated these relations and those of Slavich (1990) for their applicability in South African soils; he found Equation [22] to be more accurate for the variety of soils and situations he tested. In fact, Johnston concluded that the agreement between his measurements and those predicted using the published relationships based on Equation [22] (given in Rhoades *et al.*, 1989d) was "impressive", with "very little bias or error" and with very good correspondence (r^2 = 0.89; slope = 0.944, and intercept = 0.110). In contrast, he found the relationships of Slavich (1990), which are derived from the theoretical homogeneous depth-response functions (and as previously discussed are concluded to be inapplicable for non-homogeneous conditions), to have much poorer correspondence (slope of 0.71) and to produce a strong systematic prediction error.

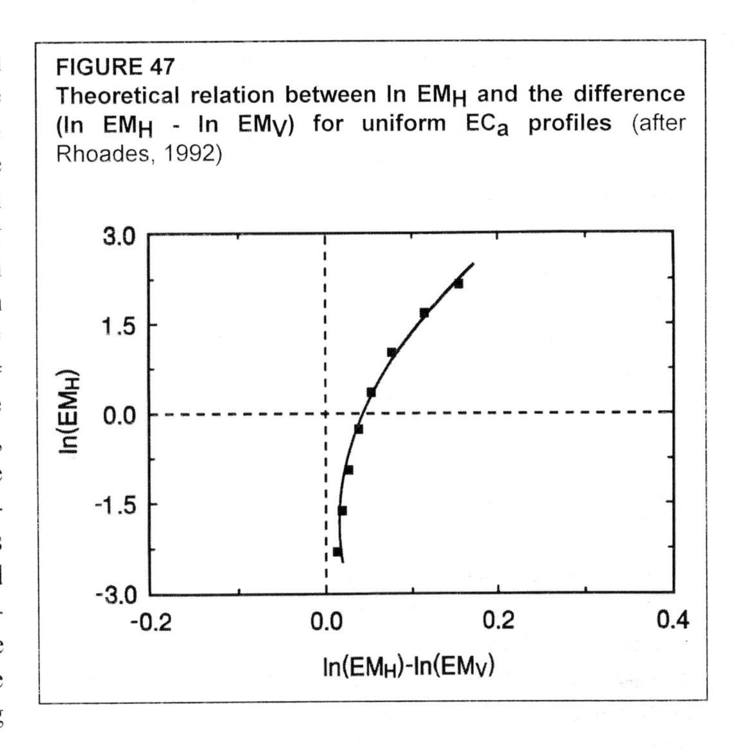

FIGURE 47
Theoretical relation between ln EM_H and the difference (ln EM_H - ln EM_V) for uniform EC_a profiles (after Rhoades, 1992)

Surprisingly, Johnston (1994) did not evaluate the following improved, more rigorous and more general relationship that was developed more recently and based on a larger and more varying set of EC_a profiles (Rhoades, 1992a) and which the author concludes to be more generally applicable and accurate than those represented in Equation [22]:

$$\ln EC_a = \beta_0 + \beta 1 \ln EM_H + \beta_3(\ln EM_H - \ln EM_V), \quad [23]$$

where β_0, β_1 and β_3 are empirical coefficients. In the earlier approach inherent in the development of Equation [22], two profile types were distinguished based on EM_V / EM_H ratios - regular ($EM_V \geq EM_H$) and inverted ($EM_H > EM_V$). Equation [22] and its manner of use have at least three deficiencies. The non-linearity that exists in the EM_H - EC_a and EM_V - EC_a relationships that occur at high values of EC_a (see Figure 46, after Corwin and Rhoades, 1990) is not taken into account; near-uniform profiles are incorporated into either regular or inverted types, and the colinearity that exists between EM_H and EM_V (see Lesch *et al.* 1992) is not taken into account. Equation [23] minimizes these deficiencies by separating soil profile types into three classes (regular, uniform and inverted), by utilizing curvilinear EM_H - EC_a and EM_V - EC_a relationships to identify the three profile types, and by using the difference (*l*n EM_H - *l*n EM_V) in place of EM_V as the second variable in the relationship, in order to minimize the colinearity problem.

The theoretical relation between *l*n EM_H and (*l*n EM_H - *l*n EM_V) **uniform** EC_a profiles is shown in Figure 47. The fitted curve ((ln EM_H - ln EM_V) = 0.04334 + 0.03058 ln EM_H + 0.00836 EM_H^2)) describes a theoretically uniform EC_a profile. Profile types may be classified based on deviation from this relation better than by the EM_V / EM_H ratio. For the practical purposes of solving Equation [23], the profile types have been classified as follows, after Rhoades (1992a): sites having values of (*l*n EM_H - *l*n EM_V) within ± 5% of the theoretical value (i.e., 0.04334 + 0.03058 *l*n EM_H + 0.00836 EM_H^2) are designated "uniform"; those with measured values > 5% of the theoretical are designated "inverted", and those with measured values < 5% of the theoretical are designated "regular". Empirically determined values of the coefficients for Equation [23] based on these classification criteria and empirical data obtained from a large number and wide variety of California soils are given in Table 1.

TABLE 1
Relationships for predicting soil electrical conductivity within soil-depth intervals from EM-38 readings[a]

Depth (cm)	Predictive Equation	n	r^2
For Regular Profiles (measured values[b] < 5% of theoretical value)			
0-30	$\ln EC_a = 0.414 + 0.985 \ln EM_H + 2.336 (\ln EM_H - \ln EM_V)$	650	0.76
30-60	$\ln EC_a = 0.836 + 1.262 \ln EM_H + 1.307 (\ln EM_H - \ln EM_V)$	626	0.75
60-90	$\ln EC_a = 0.674 + 1.089 \ln EM_H - 0.446 (\ln EM_H - \ln EM_V)$	200	0.69
For Uniform Profiles (measured values[b] within 5% of theoretical value)			
0-30	$\ln EC_a = 0.478 + 1.209 \ln EM_H + 0.411 (\ln EM_H - \ln EM_V)$	73	0.81
30-60	$\ln EC_a = 0.699 + 1.234 \ln EM_H - 0.623 (\ln EM_H - \ln EM_V)$	70	0.81
60-90	$\ln EC_a = 0.477 + 1.053 \ln EM_H - 0.691 (\ln EM_H - \ln EM_V)$	24	0.81
For Inverted Profiles (measured values[b] > 5% of theoretical values)			
0-30	$\ln EC_a = 0.626 + 1.239 \ln EM_H + 0.325 (\ln EM_H - \ln EM_V)$	56	0.91
30-60	$\ln EC_a = 0.881 + 1.216 \ln EM_H - 1.318 (\ln EM_H - \ln EM_V)$	55	0.81
60-90	$\ln EC_a = 0.563 + 1.206 \ln EM_H - 1.641 (\ln EM_H - \ln EM_V)$	21	0.91

[a] Predictions are based on measurements made with the EM-38 sensor placed on the ground in the horizontal (EM_H) and vertical (EM_V) configurations. [b] Comparing measured values of ($\ln EM_H - \ln EM_V$) with the theoretical value = ($0.04334 + 0.03058 \ln EM_H + 0.00836 EM_H^2$).

As mentioned earlier, the immediate topsoil is often dry and loose when measurements of EC_a need to be made. This has been viewed by some as a "problem" and disadvantage of the four-electrode method compared to the EM method, because of the poor electrode contact that occurs with the former method under such conditions and the lack of need for contact with the latter method (Johnston, 1994). However, as explained earlier, ***no useful information about salinity can be made from EC_a measurements made upon dry soil*** (dry soil behaves essentially as an insulator). Hence, when these measurements are made upon a soil having dry/loose surface mulch, the electrodes should be pushed through the "insulating" depth and into the moist soil to a depth of 25-75 mm (the minimum required for the electrode spacing, as explained earlier). In this manner, the dry/loose mulch does not affect or enter into the measurement of EC_a; hence, no attempt should be made to infer salinity in the depth of "bypassed" soil, nor to include it in any calibration relation. When the procedure is followed as just explained, which has always been the case with the author, there is no "problem" or disadvantage in the use of the four-electrode sensor compared to the EM-38 sensor. To the contrary, the "problem" and limitation more often occurs with use of the EM sensor. Because no contact is required with this sensor, many users simply place the unit on top of the dry/loose soil and read and interpret/calibrate it as if the "insulating" layer contributes to the reading. Thus, the EM-38 reading made in this manner includes an error, especially the EM_H reading, that is proportional to the depth of the dry layer times its weighting contribution. If the dry layer of soil is included in the calibration of the EM unit it will be in error; if the user attempts to infer the salinity in this layer from the EM-reading, it will be in error. It is inappropriate to include the dry layer in a depth-increment sample that includes moist soil. The errors created can be substantial, especially where salts are concentrated in the solid form by evaporation in the near-surface soil. Whenever feasible, the dry/loose soil should be scraped away from the measurement site before positioning the EM-38 sensor and measuring EC_a^*. This has been the routine practice of the author and his collaborators and is included in the calibrations that they have reported. It takes more time and effort to use the EM sensor in this more appropriate manner, compared to the four-electrode unit, because the dry/loose soil must be removed for the former sensor but not the latter unit. With the four-electrode unit you simply push the electrodes through the dry/loose layer; one can easily feel when moist soil is encountered and the depth of effective insertion is simply referenced from this point.

Another practice routinely used by the author and collaborators involving the EM-38 for about the last 6-7 years, is to position the axis-centers of the magnetic coils 100 mm above the soil surface at the time EC_a^* measurements are made. This practice is included in the most recent calibrations reported by Rhoades (1992a). This practice was instituted for several reasons. One is the observation made in the study of the EM-38 response to depth-varying EC_a distributions (Rhoades *et al.*, 1990b) that the reading of EM_H was higher (near a maximum) when the sensor was held at a height of 100 mm above ground than when placed on the ground. When questioned about the reason for this phenomenon, the manufacturer (McNeill, 1990) could offer no physical explanation for this phenomenon but confirmed that he had also observed it. Another reason was to keep the instrument clean by avoiding contact with muddy or dusty soil. A device was built to permit the sensor to be positioned so that the coil-axis would be located 100 mm above the soil during both EM_H and EM_V measurements. It is shown in Annex 7. This device is also used advantageously to scrape away the dry/loose surface soil before it is positioned on the ground; it is also used to "level" the EM-38 sensor. The EM-38 readings will vary some as the sensor is tilted with either the transmitter or receiver end held higher than the other. The third reason was that some height was needed to clear the clods and rough surface that exist in many fields when the sensor was incorporated into the mobilized system of measurement described earlier (Figure 41). The advantage of measurements taken at a height of 100 mm is that essentially maximum readings (depths of signal "penetration") can be achieved while avoiding contact of the sensor with the soil. It is not practical to remove the dry/loose surface layer of soil when EM-38 measurements are made with the automated/mobilized system described earlier. The associated error is minimized using a stochastic, field calibration method that creates calibrations for the specific field conditions and methods of measurement. This stochastic-method is explained later.

Small-volume Measurements

Sometimes information is desired about the levels of salinity within small, localized volumes of the soil, such as that within different sections of the seedbed or under the furrows, and about the distribution of salinity within the rootzone. For such uses/needs, the insertion four-electrode EC-probe developed by Rhoades and van Schilfgaarde (1976), and commercialized by Martek Instruments[1], Eijkelkamp Agrisearch Equipment[1] and Elico Limited[1], is recommended. The insertion EC-probe available from Eijkelkamp Agrisearch Equipment is a mechanically constructed unit consisting of four annular-ring electrodes separated by insulators directly patterned after that of Rhoades and van Schilfgaarde (1976). The construction details of the latter probe are given in the Annex 8 for those who might wish to construct their own. In the standard-sized "Martek" probe (see Figure 37), the four annular electrode-rings are molded into a plastic matrix that is slightly tapered so that it can be inserted into a hole made to the desired depth with a Lord[1]- or Oakfield[1]-type coring tube (or one of similar diameter, ~25 mm). The smaller-sized probe (so-called "bedding" probe, see Figure 37), can be simply pushed into the soft upper-soil to the desired depth. In either sized unit, the probe is attached to a shaft (handle) through which the electrical leads are passed and connected to a meter. Burial type units are also commercially available in which the leads from the probe are brought to the soil surface (see Figure 38). The original burial unit developed by Rhoades (1979b) is simple and cheap to construct, as shown in Annex 4. A multiple-depth version of the four-electrode probe has been built by Nadler *et al.* (1982). The volume of sample under measurement with any of these probe-sensors can be varied by changing the spacing between the current electrodes and the over-all diameter of the probe. The standard-sized, Martek SCT Probe, has a spacing of 6.5 cm between outside electrodes and measures a soil volume of about 2350 cm^3. The Martek "bedding" probe measures a soil volume of about 25 cm^3. Other four-electrode cells and units have been designed for other purposes including: making measurements on undisturbed soil-cores (Rhoades *et al.*, 1977), making measurements at variable water contents (Rhoades *et al.*, 1976; Bottraud and Rhoades, 1985a, 1985b; Johnston, 1994) and making measurements in soil columns (Shainberg *et al.*, 1980; Bottraud and Rhoades, 1985b). Examples of some of these units are shown in Annex 5. Very small, fixed-array units have also been constructed to measure EC_a along the wall of exposed soil profiles and in very shallow depths of soil surfaces (an example is shown in Annex 5). The possibilities are numerous and four-electrode units are easily made for such specialty purposes/studies.

When using meters which display resistance, EC_a in dS/m is calculated, for any of these probes, as:

$$EC_a = k f_t / R_t , \quad [24]$$

where k is an empirically determined geometry constant (cell constant) established for the probe in units of 1000 cm^{-1}, R_t is the resistance in ohms at the field temperature, and f_t is the factor used to adjust the reading to a reference temperature of 25° C (see Equation [2]). With the Martek unit, values of EC_a are given either at field temperature or at 25° C. Since the time response of this thermistor is slow, because it is embodied in the probe, it is usually more convenient to use an electronic temperature sensor to measure soil temperature. The author has used a soil temperature probe sold by the Wahl[1] Company for more than ten years with very good results. In normal field applications, EC_a readings are generally made in the uncorrected temperature mode and the temperature distribution throughout several soil profiles in the field is determined either with the temperature read-out of the Martek unit or, preferably, with a faster electronic temperature sensor. Subsequently the EC_a readings made at field temperature are converted to 25° C values using Equations [2] and [24].

Procedures for Interpreting Soil Salinity

Soil salinity, in terms of either EC_w or EC_e, can be determined in the field from measurements of bulk soil electrical conductivity by essentially one of three ways. Each has its own advantages and disadvantages. These alternative ways will now be described.

"Specific Field or Soil-type Calibration" Technique

Soil salinity (EC_w or EC_e) can be determined from the measurement of EC_a, or EC_a^*, made at approximately a reference soil water content using a calibration either established or predicted for the particular field or soil in question. Such calibrations are essentially applications of Equation [11]. Such linear calibration relations have been reported by Rhoades and Ingvalson (1971), Halvorson and Rhoades (1974), Rhoades (1976), Rhoades *et al.* (1977), Halvorson *et al.* (1977), Rhoades (1981), Rhoades and Corwin (1981), Cameron *et al.* (1981), Corwin and Rhoades (1982), Williams and Baker (1982), and by Slavich and Read (1983). These "pioneering" findings provided the impetus to the use of four-electrode and EM instruments to survey soil salinity that has occurred since then. Numerous satisfactory field calibrations (r^2 > 0.9) have been obtained using this empirical technique for many areas, fields and soil types around the world and successfully used to diagnose and map soil salinity. Examples of such

calibrations were given earlier (see Figures 19 and 20). Other examples are shown in Figures 48 and 49 for representative soils of the Northern Great Plains region of the United States. Various methods for establishing these types of calibrations are described in Annex 1.

Since water content affects soil electrical conductivity somewhat, as well as the relationships between EC_a and soil salinity, determinations of EC_a and the salinity-calibrations are made preferably when the soil is near field capacity. However, measurements and salinity appraisals can be made at lower water contents that exceed a certain minimum level, as discussed previously. For irrigated soils, measurements and calibrations ideally should be made after irrigation when the soil water content is at field capacity. This water content is sufficiently reproducible for such practical calibrations. Under dryland conditions, calibrations and measurements should be made in early spring, or on fallow land, in order to take advantage of the relative uniform conditions of soil water that exist then. In any case, these empirical calibrations should be established so as to apply to field soils with their natural structures, pore size distributions and water holding properties. Though Johnston (1994) has reported that he found calibrations established in the laboratory using disturbed samples were not different than those found under field conditions, this has not been the author's experience. If one desires to establish the kind of calibrations described in this section, it is recommended to do so under conditions that are as close as possible to the field conditions anticipated in their applications. This will maximize their accuracy and appropriateness. This is especially true if the EM-38 is to be used to measure EC_a, since its depth-response function will vary with the distribution and magnitude of EC_a in the profile, the presence of a shallow, saline water table, and certain other soil properties.

FIGURE 48
Relationship between soil electrical conductivity (EC_a), as determined with different interelectrode spacings and measured average soil salinity (EC_e) for a glacial-till soil in Montana, USA (after Halvorson and Rhoades, 1977)

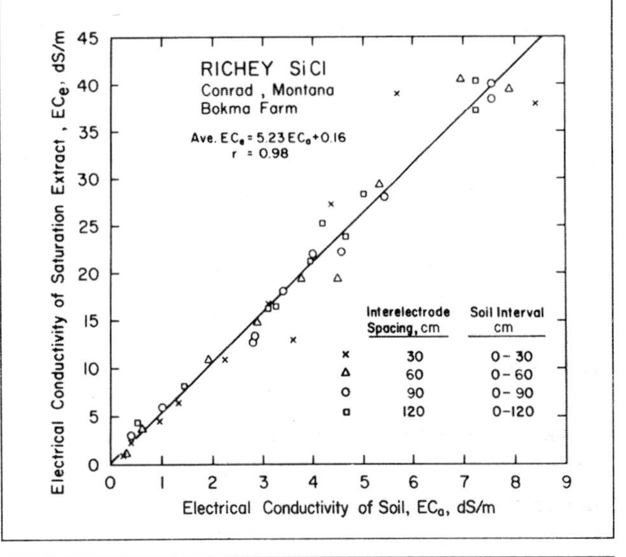

FIGURE 49
Relationships between soil electrical conductivity (EC_a) and salinity (expressed as EC_e) for representative soil types of the northern Great Plains, USA (after Rhoades and Halvorson 1977)

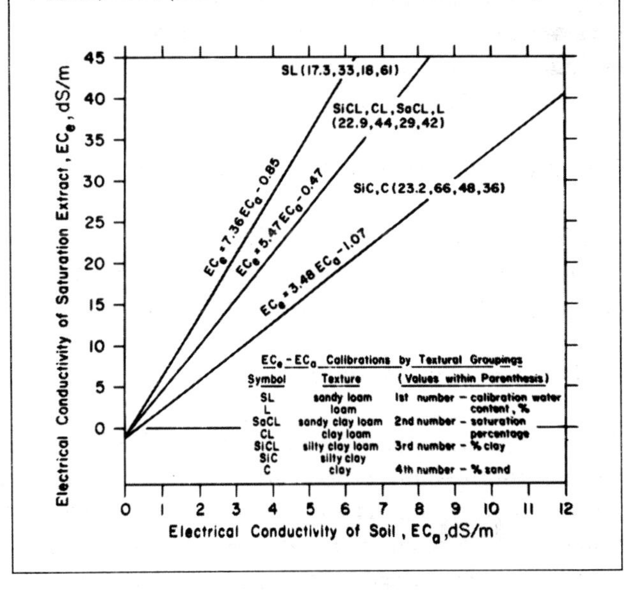

Given the value of EC_w at field capacity (or EC_e), one can readily estimate EC_w at lower field water contents using Equation [8]; the values of EC_w (or osmotic potential) occurring over the irrigation cycle can then be estimated and used to predict crop response to varying irrigation management. Of course, this estimation ignores the precipitation of salt that occurs with the reduction in water content. If more accurate estimates are required, corrections can be made for this latter process, as well as for some others, using computer models as discussed earlier. For many practical applications, the "compensation" phenomenon (implied in Equation [8] and discussed earlier) precludes the need to measure EC_w per se, or to be too concerned about measurements of EC_a having to be made exactly at calibration water content.

The measurements or bulk soil electrical conductivity may be made using a four-electrode sensor or an EM sensor. The only difference is that the latter measurement provides a more depth-weighted value of EC_a (i.e. EC_a^*). If the soil is essentially uniform in EC_a and texture, both sensors will read the same, as shown or implied by the results of Rhoades and Ingvalson (1971) and Rhoades and Corwin (1981), and their salinity calibrations (i.e., Equation [11] relationships) will be the same. That is to say, one should obtain essentially the same linear relationship between soil salinity and either EC_a, or EC_a^*, irrespective of the type of sensor used for soils of uniform properties. However, one does not expect the same calibration to be obtained from the different sensors when EC_a varies substantially with depth in the soil profile. This is implied by the finding of Corwin and Rhoades (1984) which showed that the depth-weighting response of the EM-38 sensor was different for soils whose EC_a values increased with depth compared to those whose EC_a values decreased with depth. It was conclusively shown in the study of Rhoades *et al.* (1990b) that the EM-38 depth-response relation varies with the magnitude and distribution of EC_a within the 0-2 m depth of soil. This phenomenon has important implications with respect to deciding whether it is preferable to estimate soil salinity directly from the EM readings (i.e., to establish EC_e = f (EM) relationships) or to first estimate EC_a values from a sequence of EM readings and then to estimate salinity from EC_a (i.e., to establish EC_e = f (EC_a) = f (EM) relations). The advantages and disadvantages of the two approaches will now be discussed.

Because calibrations vary with soil type and because many soils, especially alluvial arid land soils, have strata in their profiles which vary in texture and water content, it is obvious that no single EC_e (or $EC_{1:5}$) = f (EC_a, or EC_a^*) relationship can be established for such a non-uniform soil which will apply to all of its different layers/strata/depth-increments. It is often important to know the salinity distribution that exists within the various depths and strata of a root zone; for a variety of obvious reasons, knowing the mean level is not sufficient. It was for these reasons that the author and his collaborators, from the very beginning in their use of the surface-array four-electrode and EM-38 sensors to determine soil salinity, decided to develop ways to estimate the EC_a values within different soil-depth intervals (from a succession of fixed-array four-electrode and EM-38 sensor readings) and, from these values and knowledge of the soil-textural properties within the various depths, to estimate (using texture-based calibrations) the salinity level for each significant region of the soil profile. The earlier means developed for predicting depth distributions of EC_a within soil profiles from a succession of EM-38 readings has already been described; analogous means using a sequence of surface-array four-electrode readings will be described later. Also described later are the newer methods that the author and collaborators, have developed for calibrating these sensors so as to able to predict soil salinity within various depth-increments of the soil profile. Of course, when the sensor measurements are made with an insertion EC-probe, one obtains directly the EC_a within each specific depth of interest (an advantage in accuracy that is associated with this sensor). Other users of the EM-38 sensor, especially those in Canada, have chosen to correlate the EM-

38 readings with either mean profile salinity or the salinity weighted by depth in accordance with the uniform-soil depth-response nature of the sensor (Cameron *et al.*, 1981; Wollenhaupt *et al.*, 1986; McKenzie *et al.*, 1989; Slavich and Peterson, 1990). The author believes that they did this for three reasons: 1) the soils they work with are not highly stratified (they are mostly glacial-till soils, in the case of Canada), 2) their primary objective has been to map general and gross conditions of salinity, and 3) they did not have a good method for estimating EC_a by soil depth from their EM-38 readings. Johnston (1994) concluded that these latter approaches, as well as his method for estimating the mean profile salinity value from the mean ($EM_H + EM_V$) reading, were more accurate and simpler to use, hence preferable, compared to approaches which first estimate EC_a and, in turn, EC_e. He based this conclusion mostly on the **assumption** that a one-step procedure would entail less error than a two-step one and on the results of a limited test made on a relatively small number of South African soils. This assumption/premise is not inherently valid. Nor is the test convincing, since it was based on estimates of **average profile** values. For the reasons given above, different calibrations are obviously required for the different textural-layers of stratified soils. Thus, the author recommends that, where it is useful to know the salinity distribution in the root zone, the surface array four-electrode or EM-38 sensor readings be converted to their depth-increment EC_a values and these latter values be used to estimate salinity for the various important soil-depth increments of the root zone, or that direct regression relations be established between the sensor readings and the salinity levels for each important depth-increment/strata of the root zone (a stochastic method for obtaining the latter calibrations is given later). Of course, if only simple maps of gross spatial differences in salinity are needed for general characterization purposes, then the "mean" or "weighted" calibration approaches may be suitable and used.

Information about salinity within discrete soil-depth intervals can be obtained by one of three methods: 1) measurements of EC_a can be made directly within the desired depth-interval(s) using an insertion four-electrode probe (see Figure 37), 2) the EC_a values within different depth intervals can be estimated from a sequence of variably-spaced surface-array four-electrode readings (using methods described below), or from variably configured EM-38 readings (using methods described earlier for the EM-38 sensor based on Equation [23], Figure 47 and Tables 1 and 3) the EC_a values within different depth intervals can be estimated from a sequence of variably-spaced surface-array four-electrode readings and/or variably configured EM-38 readings using directly established depth-specific sensor-calibrations. The latter stochastic method is described in a later section. Of course, more accurate results can be obtained from the direct measurements of EC_a made within each depth-interval using an insertion EC-probe (Rhoades and van Schilfgaarde, 1976).

EC_a values within various soil depth-intervals, hereafter designated by EC_x , can be estimated from the sequence of EC_a values obtained with a surface-array of electrodes and successively increasing current-electrode spacings using the following relation:

$$EC_{ai - (ai-1)} = EC_x = [(EC_a \bullet a_i) - (EC_{ai-1} \bullet a_{i-1})] / (a_i - a_{i-1}) , \qquad [25]$$

where a_i represents the depth of measurement and a_{i-1} represents the previous depth of measurement. The conventional use of this equation is based on the following assumptions: the depth to which conductivity is measured is equal to the one-third of the spacing between current-electrodes (or the space between each pair of electrodes when configured in the Wenner-array) and that the stack of soil electrical resistances of a sequence of "stacked" soil layers behave analogous to resistors in parallel (Barnes, 1954). Some good results have been obtained using this approximation (Halvorson and Rhoades, 1974; Rhoades and van Schilfgaarde, 1976), as shown in Figure 50.

Alternatively, the EC_a value within a particular depth-interval can also be estimated from a succession of EM-38 readings made at different heights and coil-orientations, as explained earlier. An example of such predictions made using the earlier procedures (Rhoades and Corwin, 1981; Corwin and Rhoades, 1982) is shown in Figure 51. More generally accurate predictions can be made using Figure 47 and the relations given in Table 1, based on Equation [23].

"EC-Model" Technique

In attempts to apply "generic" soil/field-type models of the kind described above to areas mapped as the same soil-type, it was found that the estimates of salinity were sometimes not sufficiently accurate because of the substantial variability in soil properties that existed within the mapping unit. In other words, areas depicted in soil survey maps as homogeneous soil-types were found to vary considerably in soil-type within the mapping- unit. It was decided that accurate estimates of salinity for such conditions (they are not unusual; fields can be quite variable in soil texture) would require an examination of the soil profile at each measurement site and the application of an appropriate calibration for each different type of soil encountered. The following technique for determining soil salinity from EC_a was the outcome of this experience and decision. Essentially, it amounts to the practical application of the EC-model described in Chapter 3, in which field-estimates of the percent clay content

FIGURE 50
Relationship between EC_x, as calculated from Equation [25], and soil salinity (expressed as EC_e), for soil-depth intervals of 0-30, 30-60, 60-90, and 90-120 cm for a glacial-till soil in Montana, USA (after Rhoades and Halvorson, 1977)

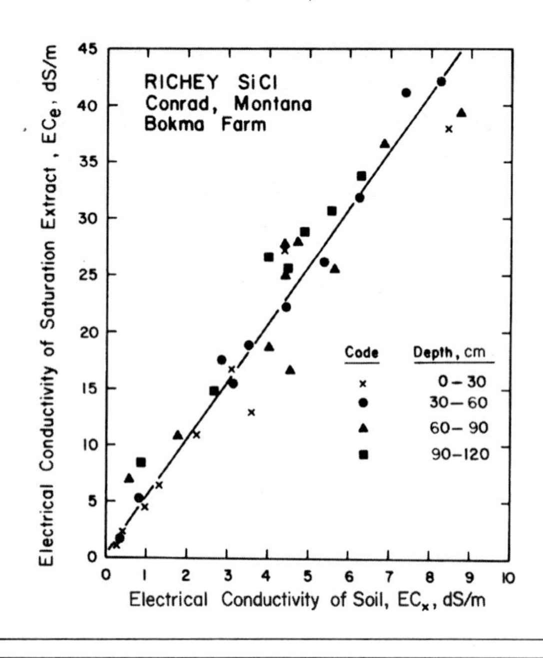

FIGURE 51
Graphs of measured and calculated (by three different methods) EC_a-depth profiles for three California USA sites (after Corwin and Rhoades, 1982)

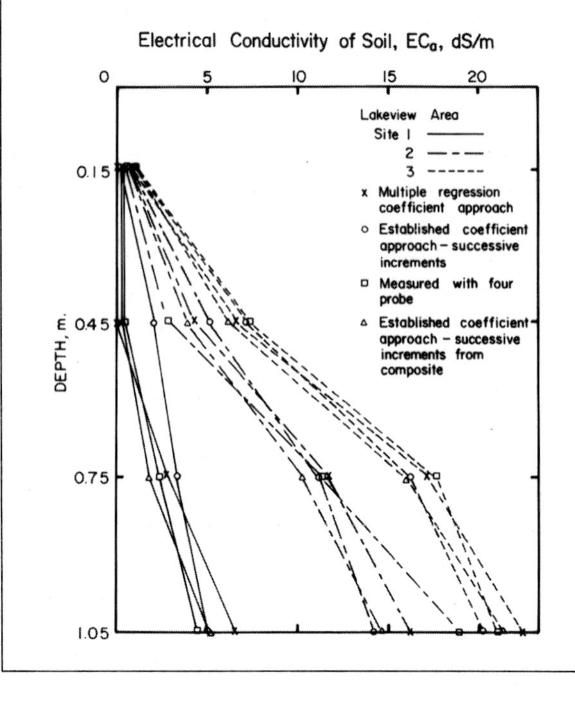

of the soil and its percent water content relative to field-capacity (for each site and depth of interest) are used along with empirically determined relations to obtain the required model-parameters needed to solve Equation [5] for salinity, given knowledge (measurement) of EC_a. These estimates of percent clay and relative water content are simply made in the field by "feel". These estimates are deemed sufficiently accurate for practical needs, as shown in Rhoades et al. (1989c, 1990a).

Thus, in this approach, soil salinity is determined for each different condition of soil-type and water content, provided the latter is in excess of the threshold value, encountered at each survey location from the solutions of Equations [5] or [6], [7] and [8] (i.e., the bulk soil electrical conductivity model and the relation between EC_w and EC_e) using measurement(s) of EC_a, estimates of soil clay percentage and percent water content relative to field capacity and the following empirical relations to estimate ρ_b, Θ_s, Θ_{ws}, Θ_{fc} and EC_s:

$$SP = 0.76\ (\%C) + 27.25\ , \qquad [26]$$

$$\rho_b = 1.73 - 0.0067\ SP\ , \qquad [27]$$

$$\Theta_s = \rho_b / 2.65\ , \qquad [28]$$

$$\Theta_{fc} = SP(\rho_b / 200)\ , \qquad [29]$$

$$\Theta_w = \Theta_{fc}\ (FC/100)\ , \qquad [30]$$

$$\Theta_{ws} = 0.639\ \Theta_w + 0.011\ , \qquad [31]$$

and

$$EC_s = 0.019\ SP - 0.434\ , \qquad [32]$$

where %C is clay percentage as estimated by "feel" methods, Θ_{fc} is the volumetric water content at field capacity, and FC is the percent water content of the soil relative to that at field capacity as estimated by "feel".

Given the above assumptions, estimates and measurement of EC_a, EC_w is calculated from the solution of Equation [7]. Then EC_e is determined from Equation [8], assuming that $EC_{wc} \cong EC_{ws}$ and, therefore, that $(EC_w\ \Theta_w) \cong (EC_{wc}\ \Theta_{wc} + EC_{ws}\ \Theta_{ws})$. These calculations can be made simple using a programmable pocket calculator; alternatively, EC_e can be obtained graphically using Figure 52, after Rhoades (1990b) and Rhoades and Miyamoto (1990). Examples of the successful use of this technique are given later.

Sensitivity analyses and results of field tests have shown that the estimates and assumptions described above are generally adequate for practical salinity appraisal purposes of typical mineral, arid-land soils (Rhoades *et al.* 1989c and 1990a); i.e., that EC_e can be estimated in the field sufficiently accurately for most salinity appraisal purposes from the accurate measurement of EC_a and reasonable field estimates of %C and FC made by "feel". For organic soils, or soils of very different mineralogy or magnetic properties, these estimates may be inappropriate. For such soils, appropriate estimating procedures will have to be developed using analogous techniques to those used by Rhoades *et al.* (1989a). The accuracy requirements of these estimates may be evaluated using the relations given in Rhoades *et al.* (1989c).

FIGURE 52

Relationships between electrical conductivity of bulk soil (EC_a), electrical conductivity of saturated-paste extract (EC_e), relative soil water content as percent of field-capacity, and soil clay content (% clay), for representative arid-land soils (after Rhoades and Miyamoto, 1990)

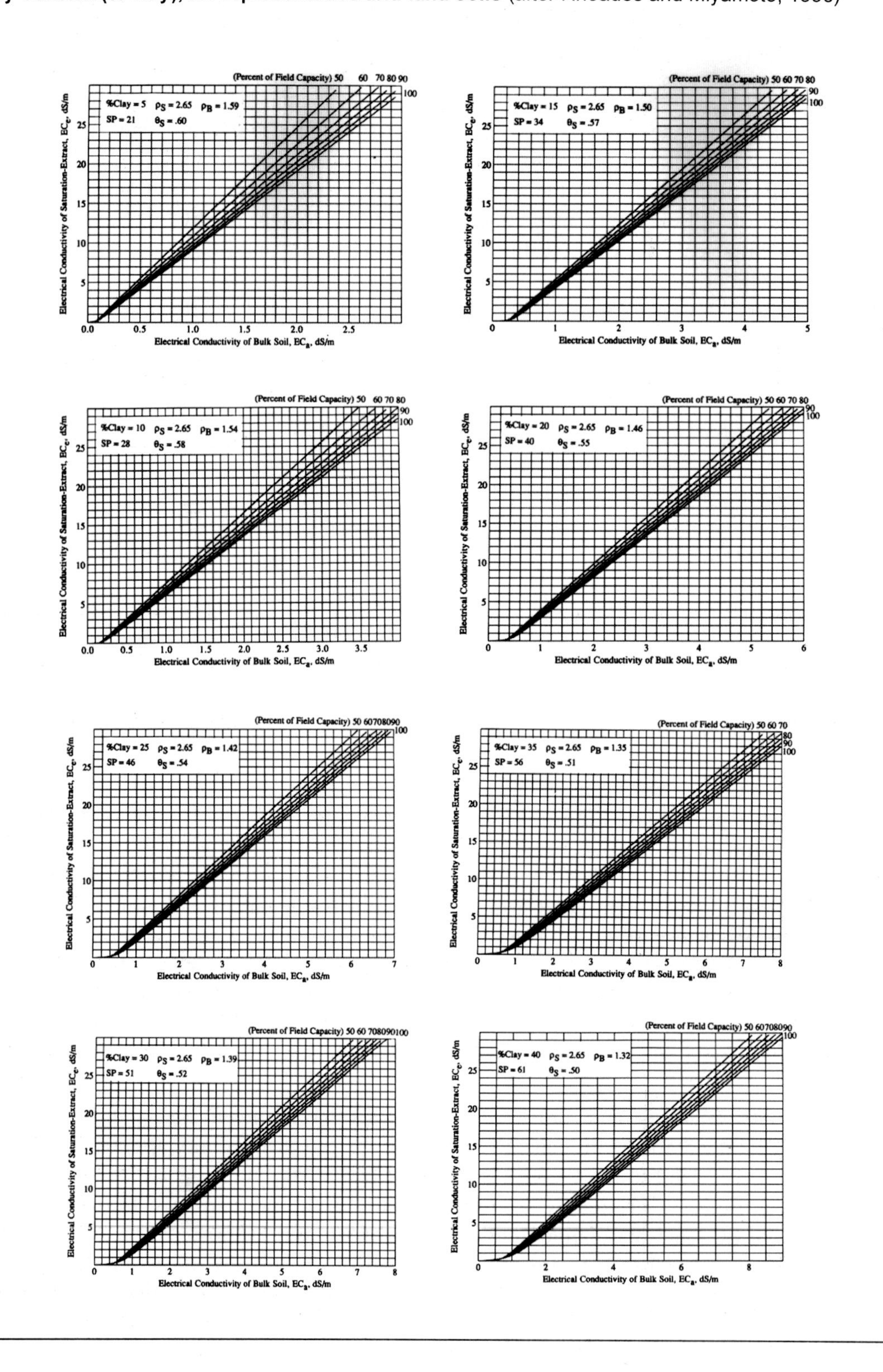

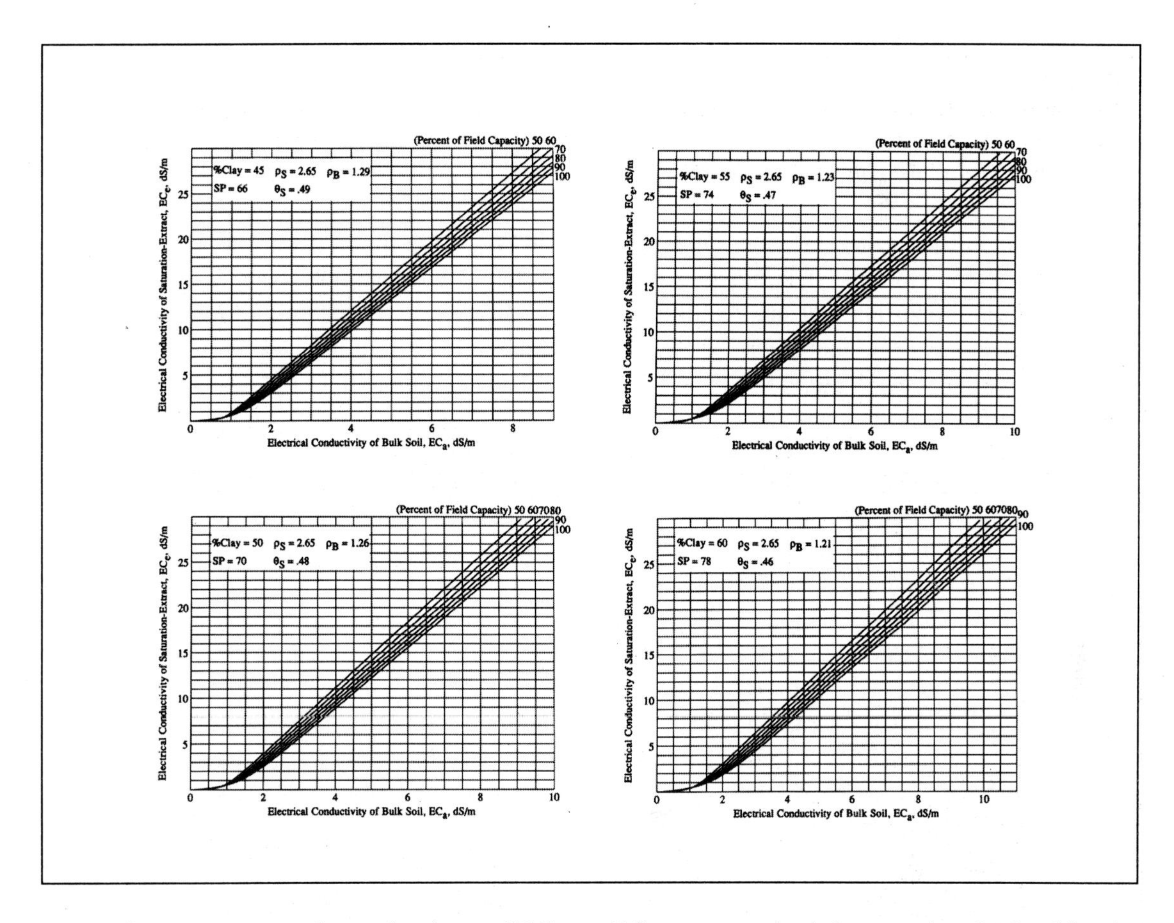

If more accurate determinations of EC_e, or EC_w, are required than can be obtained by the estimation procedures described above, then quantitative measurements of Θ_w, EC_s, ρ_b, etc. should be made using appropriate methods and used in place of the above-described estimates.

The major advantage of the "EC-model/field-estimates" method described above is that it accounts for the site-to-site variabilities in soil properties (clay and water content in particular) that can occur within individual fields, mapping units, or other areas of interest. Essentially, it generates a specific calibration between EC_a and EC_e, or EC_w, for the particular soil condition encountered at each site of EC_a measurement in the field (area) under evaluation. Such "specific" calibrations are generally more accurate than are the "average soil-type" calibrations when applied to an area of field size or larger which is assumed to be the same as the calibration soil-type (Rhoades *et al.* 1990a). Field tests of this method have shown it to be sufficiently accurate for the practical purposes of salinity diagnosis and mapping, to be faster than conventional soil sampling and laboratory methods (measurement of EC_e per se, either directly or as estimated from EC_p) and to be generally more accurate than the "soil-type" calibration (Rhoades *et al.* 1990a; Annex 1). Another advantage of this approach is that it saves the time involved in establishing calibrations for each of the different kinds of soils found in the survey area. The major disadvantage of the method is that one must estimate (by feel) the clay percentage and relative water content of the soil at each site and for each depth of measurement. This requires that the soil be probed at every measurement site and that time be taken to make these estimates. Of course, when one collects sensor readings at any "unexamined "site and estimates the corresponding levels of salinity, especially within the various depth increments of the root zone, by one of the other approaches, there will always be uncertainty about the properties of the soil within the profile at that site and, hence, about the appropriateness/applicability/accuracy of the predictions. As will be discussed in more detail later, the combination of the EC-model with the stochastic field-calibration approach, which is described next, is very appealing.

Stochastic Field-Calibration Technique

As explained earlier, when measurements of EC_a, or EC_a^*, are made using the mobilized and automated sensor systems, it is not possible to simultaneously estimate or measure the other secondary soil properties that are required to use the "EC-Model" technique described in the preceding section. For that reason, the "stochastic field-calibration" technique that was mentioned earlier was developed. This latter method is also applicable to hand-held sensors and is more rigorous than the "generic field or soil-type calibration" technique. This stochastic field-calibration technique is essentially a statistical/ground-truthing approach in which a predictive (regression) relationship between EC_a, or EC_a^*, and EC_e (or EC_w) is established and, subsequently, used to determine salinity from sensor measurements in a calibrated area of land that is relatively homogeneous with respect to soil conditions other than salinity. In this approach, spatial regression modeling techniques are usually preferable to the classical geostatistical modeling techniques, since the former require less calibration data and are typically easier to estimate (Lesch *et al.*, 1995a, 1995b).

The numerous sensor readings of EC_a^* are obtained within the sampling area (usually a field) under evaluation on a uniform (centric systematic) grid basis. Based on the observed field pattern of EC_a, or EC_a^*, readings, a relatively small number of sensor measurement sites are chosen for soil sampling using a statistical model/procedure, which is described in more detail later. Soil samples are collected at these sites and their salinities are determined by any accepted method of salinity appraisal (the EC-paste method of Rhoades *et al.* 1989b is recommended for this purpose). A multiple linear regression relation is then established between the EC_a and/or EC_a^* readings, the measured soil salinities (EC_e, EC_w, or some other expression of salinity), and the x/y coordinates for each soil depth of interest. Such a spatial regression model can be written in matrix notation as:

$$Y = X\beta_1 + W\beta_2 + \varepsilon, \qquad [33]$$

where Y represents the vector of log transformed soil salinity values, X represents a matrix of log transformed and de-correlated sensor readings, W represents a trend surface matrix based on the spatial coordinates of the measurement sites, and ε represents a random error component. For the fairly typical case where W is a 1st order trend surface matrix, equation [33] becomes:

$$\log(EC_e) = \beta_0 + \beta_1 \log(EM_H) + \beta_2 [\log(EM_H) - \log(EM_V)] + \beta_3 \log(EC_{4p}) + \beta_4 X + \beta_5 Y + \varepsilon \quad [34]$$

where EM_V and EM_H are as previously defined, EC_{4p} refers to EC_a as determined from the four-electrode sensors, (X,Y) refer to the spatial-coordinates of the measurement sites, β represents the regression-fitted parameter estimates, and ε is the random error component. The resulting field-specific relation is subsequently used to predict the salinities at the vast number of unsampled sites/depths in the area where the remainder of the EM-38 and four-electrode sensor measurements were made. This "single-step" method eliminates the need to first convert EM-38 (essentially EC_a^*) readings to equivalent EC_a values within a particular soil depth using general relations (such as those in Table 1 which are based on Equation [23]) and thence to EC_e, as is required by the "EC-Model" technique.

Software (Estimated Salinity Assessment Programme, ESAP) has been developed to determine the regression model and to prepare a map of the predicted salinity pattern for a surveyed field; the procedures for use of this software are described, with examples, in a User Manual (Lesch *et al.*, 1995c). The ESAP software also provides an algorithm to determine the numbers and locations of sites in a surveyed field to be soil sampled for calibration purposes; this is discussed more in a following section. This software is available from the US Salinity Laboratory. A "windows-based" version with various "user-friendly" features is now under development; it should be available by the time this paper is published.

Experimental results show that this method works very well for fields/landscapes that are relatively homogeneous in all factors affecting EC_a conditions other than salinity, such as individual fields uniformly managed or sections of natural landscapes that have similar soil types and properties (certain dryland landscapes for example). This approach substitutes easily acquired EC-sensor field measurements for the more difficulty carried out procedures of soil sampling and laboratory analysis. It very substantially reduces the number of soil samples required by traditional methods to accurately and intensively map the spatial salinity patterns within fields, as well as the overall cost. Larger areas of land can be mapped by joining adjacent areas on a field by field basis. This method is more practical than those based on conventional geostatistical procedures, such as those traditionally used for salinity mapping purposes (i.e., Webster, 1985, 1989), because it reduces the intensive soil sampling generally needed to obtain the accurate variogram estimates required in these latter procedures (unpublished data). The major limitation of the method is the requirement that the fields be under relatively uniform in management and that soil water, bulk density, and clay content be reasonably homogeneous. If needed however, larger fields (or areas) can be subdivided into smaller more homogeneous units and the method applied analogously to each sufficiently homogeneous subunit. Alternatively, additional practical measurements besides EC_a, such as location coordinates, elevation, etc., can be made and incorporated into the regression relation (such as coefficients β_4 and β_5 do in Equation [34]) to adjust for some of the "other" factors influencing the salinity prediction (Lesch, et al. 1992). Disadvantages of the method are the need to enter the field a second time after the EC_a-sensor readings have been taken to locate the selected sample-sites and to acquire the "calibrating" soil samples. The latter locations are not difficult to establish when numbered markers have been left in the field at the sites of each EC_a measurement, or with the use of "real-time" GPS systems. This need for re-entry is not a major factor when large areas are being mapped. A soil sampling team is usually sequenced one-day after the EC_a measurement operation; the statistical calculations used to select the sampling sites can be made at field-side by another team member. This method is especially appropriate where very rapid, mobile instrumental systems are being used to intensively map large fields or areas of land.

While the use of both EM-38 and four-electrode measurements are included in Equation [34], analogous relations can be developed using just EM data or just four-electrode data. Furthermore, the relations can be developed for a single soil depth or for a series of soil-depth intervals. These empirical relations typically yield highly accurate predictions of the spatial pattern of soil salinity in fields, since the model is specifically calibrated for each field (Lesch*et al.*, 1992, 1995a,b). They are highly field-specific and can not be developed without calibration data. Such relations can also be used for monitoring purposes (i.e., testing for a change in the condition of salinity over time), provided additional soil samples are acquired at the future time (Lesch *et al.*, 1998). This use is discussed later in the context of salinity monitoring.

Since the major effort involved in this "stochastic-calibration" approach stems from the collection and analysis of the soil samples, one tries to minimize the number of calibration sites.

The location of these sites are chosen so they meet certain statistical criteria, such as the optimal estimation of the regression parameters and /or the minimization of the prediction error. An algorithm was developed for generating such a model-based sampling/calibration methodology for use with the mobilized combination-sensor, salinity assessment systems described above (but which has more general applicability) and it is described in Lesch*et al.* (1995a, 1995b) and in the Estimated Salinity Assessment Programme (ESAP) Software (Lesch *et al.*, 1995c).

The good success obtained in determining the levels and spatial patterns of soil salinity in irrigated fields and crop root zones using the various above-described methods may be seen in the following publications (Lesch *et al.*, 1992, 1995b; Rhoades, 1992b, 1993, 1994, 1996b; Rhoades *et al.*, 1997a, 1997b). Examples are given later which illustrate the typical success achieved with this stochastic field-calibration approach to measure and map salinity. While the above described approach has been found to be generally quite robust and accurate, it is not suitable if the field is very heterogeneous in soil type. For such situations, the EC-model (Equation [5]) method is advised, or the combined use of it with the stochastic-calibration method.

Comparisons of the Different Methods of Measuring Soil Salinity

Only a few direct comparisons of the various instrumental and conventional methods of measuring soil salinity have been made to date. Salinity measurements were made by four methods in a field experiment in India (Yadav *et al.*,1979), i.e., porous-matrix salinity sensor (EC_m), vacuum-cup soil water sampler (EC_v), soil samples (EC_e), and four-electrode soil conductivity sensor (EC_a; surface Wenner-array method). These investigators found a better linear correlation between EC_a and EC_e ($r = 0.93$) than between EC_e and EC_m ($r = 0.78$) or between EC_e and EC_v ($r = 0.78$). They concluded that, for purposes of diagnosing the salinities of the soils of an extensive area, the four-electrode technique is preferred because it is more rapid, simpler, and more practical. Loveday (1980) compared the four-electrode (surface Wenner-array) technique with soil sample extracts (EC_e) in a survey of 50 field sites in Australia. The water contents of the soils at the time of measurement were not generally at field capacity. He obtained relatively high correlations between EC_a and EC_e, though variance was high. He attributed this high variance to field variability factors and concluded that the four-electrode method was good for gross survey work but not accurate enough for predictive purposes. However, Loveday used generic soil-type calibrations and only two 5 cm-diameter soil samples to estimate the salinity of the relatively large volume of soil included in the Wenner measurement. One must question that such small samples represent "ground truth", and hence the appropriateness of his conclusion. ***Indeed, it has been found that the so-called salinity "ground-truth", as typically determined using small-volume soil samples, used to test the credibility of instrumental techniques of salinity appraisal are usually not very representative of the larger volume of soil involved in the instrumental measurements*** (Rhoades *et al.* 1989d, 1990a; Lesch *et al.* 1992). Loveday also concluded from his results that EC_e - EC_a calibrations found in the US by the author and collaborators were probably universally applicable to soils of similar texture. Van Hoorn (1980) compared salinities measured using extracts of soil samples with both those obtained by four-electrode surface-array and four-electrode EC-probe methods in large experimental tanks. He concluded that for survey work either the Wenner method or the four-electrode probe could be used, but that the accuracy of the latter is much greater. Nadler and Dasberg (1980) compared soil salinity measurements made in small salinized field plots using in situ ceramic porous matrix sensors, four-electrode EC-probes, a four-electrode Wenner-array, and soil sample extracts (1:1). They found good correspondence between "expected" salinity and both "soil extract" salinity and "four-electrode" salinity, but not with "porous matrix salinity sensor" salinity. They attributed the latter discrepancy to lag-time problems. They concluded that the Wenner-array method could be used more reliably under drier soil conditions than could the four-electrode probe, which requires better electrode-soil contact for accurate measurements. Johnston (1994) evaluated the suitabilities of four-electrode and EM-38 sensors for appraising soil salinity, as well as various means of their calibration. The authors has already discussed most of his findings and conclusions in previous sections of this paper. He found that the methods were suitable for practical salinity diagnosis and mapping purposes, given appropriate site-specific calibrations, and concluded that the EM-38 was more convenient and accurate. As the author has already stated, he does not agree with all of his conclusions for the various reasons previously given. While he made the most thorough and field-based evaluations of the attributes of the various sensor-methodology that is the subject of this paper, much of it was directed to the earlier less-developed procedures than those advocated herein.

The author obtained good results with the use of both four-electrode and EM-38 sensors and with all of the various methods of soil salinity assessment described in this report, not only in the numerous locations in the US, but also in other countries. He has found these sensors and techniques to be useful in varied applications including salinity diagnosis, mapping, monitoring, saline seep and water table encroachment identification, irrigation scheduling and control, leaching fraction assessment, identifying areal sources of salt-loading, evaluating drainage adequacy, evaluating irrigation-infiltration uniformity, and in developing site-specific farming plans. The author finds the porous matrix salinity sensors to be less generally useful than the sensors which measure soil electrical conductivity, because of their small sampling volume and of their substantial lag time response to changing soil salinity situations. However, for some applications they may still be the preferred technique. He also finds the TDR sensors to be less useful for salinity appraisal than the four-electrode and EM-38 sensors, because the former are less robust, more limited in their volume of measurement, less adaptable to mobilization, more limited in the range of salinity they will sense, and more time-consuming in data acquisition and interpretation. But in fact, the author recommends that the various techniques be used complementarily; the mobilized and combined EM/four-electrode system is most suitable for surveying large fields in detail to establish the larger scale spatial variability in salinity conditions and the underlying causes of it, and the four-electrode probe is more suited to acquiring detailed information of EC_a (and salinity) within various regions of the root zone, such as below the furrow, within the bed, with distance from drip emitters, etc. The fewest appropriate number of soil samples can then be taken from the different areas for detailed chemical analysis of the salinity composition, if desired, using the salinity variability information obtained with the instrumental readings and the ESAP software. This "combined-use" approach greatly facilitates the tedious, time-consuming and costly aspects of soil sampling. Whether the soil samples are reacted with water, or soil water per se is isolated from the soil for detailed analysis, is a matter of need and practicality. For practical reasons, aqueous extracts are generally used, although ideally one would prefer an analysis of the actual soil water. When an extract is to be used, it should be the one with the lowest water:soil ratio feasible. The EC-paste method has proven to be very accurate and dependable; it can be used advantageously in lieu of extracts when soil samples need to be analyzed only for salinity. The new SAR-paste method, referred to earlier, can be analogously used to diagnose and screen soil samples that need to be appraised, especially in the field, for sodicity problems (Rhoades*et al.*, 1997c).

The equipment and procedures described and advocated herein are all undoubtedly useful for many purposes of salinity assessment and can be used in many different ways; the most appropriate will vary with the exact needs and circumstances of the user. It is for this reason that the approach taken in this report is to provide a fundamental understanding of soil electrical conductivity, of the attributes of the various sensors and methods of their calibration and interpretation, so that the various kinds of users can adapt them as needed to meet their specific applications and circumstances. Such users should always be aware of the limitations inherent in each of the alternative methods of measuring soil salinity and take them into account. The most appropriate one(s) should be used according to the specific needs and objectives of each particular situation. Again, the overall task of measurement and monitoring of soil salinity can be greatly facilitated through the combined use of the various methods. The EM and four-electrode instrumental methods should be used for most of the field characterization needs; laboratory analyses can then be carried out on only the minimum appropriate number of soil samples collected in accordance with the findings of the surveys made with the field-instruments. The areas requiring separate sampling are most easily determined from mobilized EM/four-electrode measurements; the depths to be sampled and numbers of samples to be taken from within each sampling area/depth are most accurately determined using the ESAP software. The most accurate salinity profile information can be determined from four-electrode probe readings and the EC-model relations.

A comparison of the costs and time involved in making salinity assessments at the field and regional scales using the different methods and equipment reviewed above is given in Chapter 5. The results are summarized in Table 26, along with a comparison of the differences in the amount of information/data provided by the different methods. Only the overall results of this evaluation will be summarized here.

The evaluation should be understood to be only relative, since the actual costs (both capital and operational) will vary from one country to another depending on differences in their labor costs and technical development. The costs are based on conservative estimates and US conditions; if there is bias in this evaluation, it is made intentionally in the favor of conventional soil-sampling and laboratory analysis methodology, so as not to over-promote the instrumental methods. The cost to undertake detailed field-scale surveys (for a typical 64-hectare field, 12 by 12 grid and three soil-depths) using soil samples and traditional laboratory extraction procedures is concluded to be completely cost prohibitive ($146 per ha.; 43 hours of field time). The cost can be reduced to $26 per ha. using the EC-paste method of Rhoades *et al.* (1989b), but with no savings in field time. The cost can be reduced even more with the use of field instrumentation which measures bulk soil electrical conductivity. Detailed field-scale surveys can be made using three different hand-held, instrumental approaches (two of which require the use of some soil samples for calibration purposes) at a cost savings of 96% compared to conventional method. Such surveys can be accomplished: 1) using the EM-38 at a cost of $6.50 per ha., requiring 4.4 hours of field time, 2) using a four-electrode surface-array sensor (two soil depths only) at a cost of $7 per ha., requiring 6.2 hours of field time, and 3) using the EC-probe (without the analysis of any soil samples) at a cost of $6.60 per ha., requiring 14.6 hours of field time. With the use of the mobilized instrumental systems, the costs can be lowered even more when regional assessments are undertaken ($3.24-$3.68 per ha.) and are as cost effective as the hand-held instrument methods for detailed field assessments while providing substantially more spatial information in the latter case. An added benefit of the mobilized instrumental methods, compared to conventional methods and hand-held instrumental methods, is the very substantial reduction in the field time that is required by the former methods (the required time with these methods is 2.7-3.7 hours per 64-hectare-field). It is concluded that the instrumental methods of soil salinity assessment described herein are very cost effective; they are also very time effective. As shown in this evaluation, the savings in field time is substantial with the use of the field instrumental methods. Another time savings feature of these methods, but not included in the evaluation/comparison, is the timeliness of the information/results they provide. With the mobilized systems, a detailed field-scale survey can be completed in the same day. With conventional methods, weeks to months are usually required in this regard.

Determination of Locations of Measurement and Calibration Sites

For purposes of salinity mapping and many other salinity assessment applications, the sites where EC-sensor readings or soil samples are taken must be associated with geographic location (x- and y-coordinates). If rapid methods of salinity measurement are to be used to best advantage, then an equally rapid means of determining sample site location must also be used. For this purpose, the LORAN and GPS systems used in marine and aviation navigation can be employed, where it is available, with success for certain types of salinity mapping (see Rhoades *et al.* 1990a, 1990c). The LORAN-C system, which broadcasts pulsed radio signals at a frequency of 100 kHz, is operated by the U S Coast Guard in cooperation with several other countries as an aid to marine navigation. The coverage is very good regardless of terrain and is not limited to line-of-sight transmission because of the use of the LF radio waves. The LF radio band is propagated by means of the Ground Wave, so that the radio waves closely follow the surface of the earth. The receiver calculates its' location by measuring the time delays of the received signals from three different transmitter locations and applying the principle of triangulation. Thus, the LORAN-C receiver is essentially a precise time-difference measuring instrument which processes the received information to determine a position-fix. The position-fix is given directly in terms of latitude and longitude coordinates expressed in degrees, minutes and seconds. With local calibration the repeatability of position determination can be as good as 10 meters, or better (Rhoades *et al.* 1990c). This "accuracy" is good enough for regional surveys, but not so for more detailed mapping and small-scale assessment applications.

The global positioning system (GPS) provides a more accurate and generally available means to establish sample-site positions for assessment and mapping purposes. The GPS is a satellite-based radio navigation system operated by the U S Department of Defense. It consists of 21 satellites in circular orbit at a 20,000 - km altitude and provides world-wide, 24-hour coverage. The system is analogous in concept to LORAN, but utilizes the line-of-sight reception of signals from multiple satellite-based transmitters of known position. A GPS receiver unit obtains/deciphers coded and synchronized signals emitted by several (usually at least 3) GPS satellite - transmitters in terms of time of measurement, distance from the transmitter, and position of the receiving antenna. Distances are determined by measuring the difference in time it takes the radio signal to travel between the various satellites and receiver by means of accurate, synchronized clocks contained within both the transmitters and receivers. The receiver (sample) position is calculated by "triangulating" the range distance from three or more satellites of known position. These calculations are carried out by the GPS receiver. The Global Positioning System includes five control stations evenly spaced around the earth near the equator. They track each satellite, determine their exact positions and transmit correction factors to the satellites and, in turn, back to the receivers. The suitability of the simplest/cheapest GPS for soil survey purposes has been shown by Long *et al.* (1991) to meet the accuracy requirements for detailed soil surveys (30.5-m). Accuracy is increased by averaging multiple readings taken over 10 seconds and by post-processing the data obtained by the mobile-receiver data to correct it for "drift" using analogous data collected over the same time period with a fixed-base, reference receiver (base station). Accuracy's of receiver position to within 2 to 5 m of true is made possible through use of this so-called "differential-mode" of operation (two receivers; one mobile and one stationary) and post-processing technique. Positional accuracy, under the mountainous and forested conditions of western Montana (USA), was found to average between 3 and 4 meters in the open and between 5 and 6 meters under closed forest canopies (Gerlach and Jasumback, 1989). Real time, differentially-corrected readings can be obtained using radio receivers and transmitted "corrections" from either dedicated stations or your own base station. The latter procedure is used in my mobile salinity assessment systems to establish the coordinates of EC_a measurement sites and of soil sampling locations (Rhoades, 1992a,1992b; Rhoades *et al.*, 1997d). The accuracy of this system is about 20 cm. The differential corrections can also be made in real time by incorporating a portable PC/software into the GPS system. There are numerous companies now selling this kind of equipment; the reliability and convenience features have been increasing steadily as the cost has been steadily declining (some costs are given in Chapter 5). The technology is well developed and extremely useful, in fact almost a necessity for salinity assessment techniques based on the mobilized system advocated herein.

In an earlier section, the stochastic field-calibration technique for predicting soil salinity from EC_a-sensor readings was described and advocated as an accurate and efficient means for assessing salinity at the field scale. In this method a spatial multiple linear regression model is established for the surveyed field based on intensive sensor readings and limited soil samples collected and analysed for salinity. The intent is to minimize the number of soil samples used in the calibration while assuring that the calibration is representative of the whole field. A site selection algorithm has been designed to facilitate this calibration, and also to select sites for follow-up monitoring evaluations. The algorithm selects sites that are spatially representative of the entire survey-area and simultaneously facilitates the accurate determination of the model parameters, based on rigorous geostatistical procedures. The advantage of the algorithm is that it is more cost-effective compared to conventional cokriging methods; regression models can be fitted with substantially fewer calibration sample sizes than is required with cokriging. With this algorithm, a suitable stochastic field-calibration is obtained with a small number of calibration sites ($n \sim 8$-20, depending upon the accuracy requirements of the survey) by combining the survey site information with response surface design techniques. It ensures that the selected set of calibration sites (i) is spatially representative of the entire survey-area and (ii) is suitable to permit the efficient determination of the regression equation parameters of Equation [34].

This sampling-location algorithm is provided in the ESAP software package. The details of the procedure are described in Lesch *et al.*, (1995a and 1995b). Briefly, this algorithm transforms and decorrelates the EC_a readings by a principal components analysis; it uses the transformation in conjunction with a response surface design to identify a statistically efficient set of calibration sites, and it modifies the response surface design as needed to optimize the spatial locations of the final calibration sites. An earlier version of this approach is described and illustrated in Lesch *et al.* (1992). An example illustrating the locations of the sensor readings and the calibration sites is given in Figure 58, which is discussed later.

Chapter 4

Example uses of salinity assessment technology

The following examples are intended to illustrate the utility of the instrumental salinity assessment technology described earlier. It is not intended to show how to apply the equipment to each and every kind of problem and situation; there are too many combinations in this regard. Many of the examples are based on earlier studies carried out before subsequent improvements were made in equipment, spatial statistics, software, theory/models and empirical relationships. Thus, better results than those obtained in some of these examples could most likely be obtained today using the newer improved versions of the technology, especially the mobilized systems, the EC-model technique and the stochastic field-calibration technique. Still, the simpler earlier methodology can be useful for many applications. Thus, the examples given should be viewed as instructional and as illustrative of the many ways and opportunities that exist to utilize the measurement and assessment technology and as a means to help the readers envision other possibilities and manners of utilization; the possibilities are numerous.

Diagnosis of Soil Salinity and Saline Seeps

A saline seep is an area of formerly productive non-irrigated soil that has become too wet and saline for economical crop production, as a result of the flow of saline subsurface water to the soil surface. Seeps generally develop on the lower positions of hillsides where there is a change in slope. Typically, water percolates through the soil profile located in an upslope recharge area, picks up salt in the process, is intercepted by a slowly permeable horizontal stratum, moves laterally through the relatively more permeable layer, resurfaces where the permeable stratum is truncated on a hillside, evaporates and deposits the accumulated salt, forming a saline seep. The recharge area is that area upslope from the seep (discharge area) from where the percolating water originates. Typically, excess percolation is caused by the conversion of permanent vegetation of higher net evapotranspiration to an annual crop of lower usage. Problems in combating saline-seep development are associated with diagnosing soil salinity, identifying potential saline-seep areas, determining increasing soil salinity trends in the field, and detecting the encroachment of a shallow perched water table before excessive crop damage occurs.

The levels and distributions of salinity in the landscape, especially in the soil profile, were shown to be good indicators of encroaching seep development, as well as of the likelihood of incipient crop failure. Plots of EC_a, as determined by surface-array, four-electrode measurements, versus interelectrode spacings (essentially equivalent to soil-depth in the Wenner-array) yielded distinctively shaped curves for recharge areas, encroaching shallow water tables, and seep areas, as shown in Figure 53 (after Halvorson and Rhoades, 1974). The curves for seeps showed a sharp decline in EC_a with depth indicative of salts being carried upward to the soil surface from the shallow water table and their accumula-tion at the surface by evaporation. In contrast, curves for recharge areas showed a gradual increase in EC_a with depth and then a steady level with further depth corresponding to the leaching of salt in the soil profile and a net downward flux of water. Curves in areas that were not yet excessively salinized but which had water tables approaching critical depths displayed a level of EC_a (salinity) that was significantly higher than typical of the normal regional soil. These areas also displayed an initial sharp increase in EC_a with depth (to about 0.5 m) and then a gradual decrease with further depth corresponding to the interaction of leaching of salts in the near-surface soil and the upward flow of saline shallow groundwater into the soil profile. These salinity levels and distributions are distinctive and readily interpretable in terms of the processes of leaching and/or drainage.

FIGURE 53
Relationship between electrical conductivity of bulk soil (EC_a) and "Wenner-array" interelectrode spacing (approximately equivalent to soil depth) for a saline seep, an encroaching saline seep site, and an unaffected site for glacial-till soil in Montana, USA (after Rhoades and Halvorson, 1977)

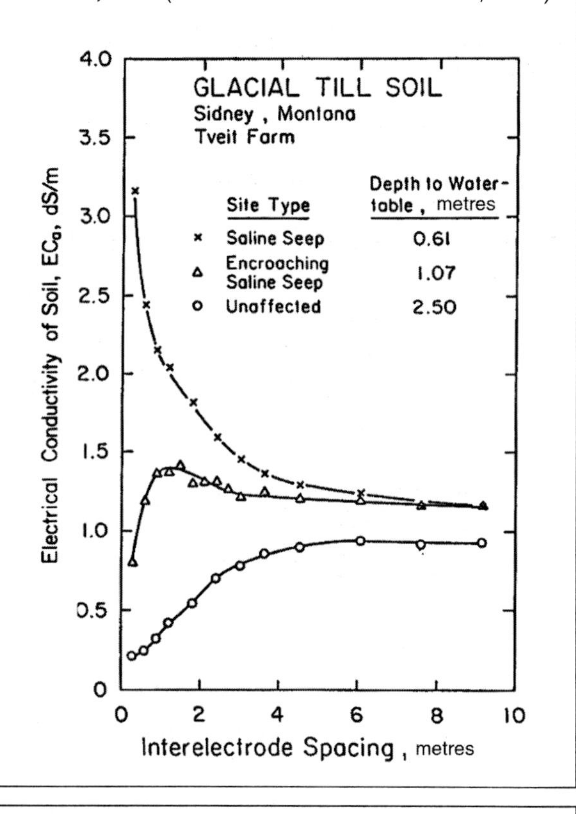

The presence of a water table within the critical depth (capable of contributing salts to the root zone by capillary flow) in the soils of the Northern Great Plains (USA) where saline seeps occur was found to be detectable from the level of salinity (or EC_a) in the topsoil (0-30 cm depth), as is illustrated in Figure 54 and in Table 2.

FIGURE 54
Relationship between electrical conductivity of bulk soil (EC_a) in the 0-30 cm. depth- increment and depth to water table in typical glacial-till soils of Montana, USA (after Halvorson and Rhoades, 1974)

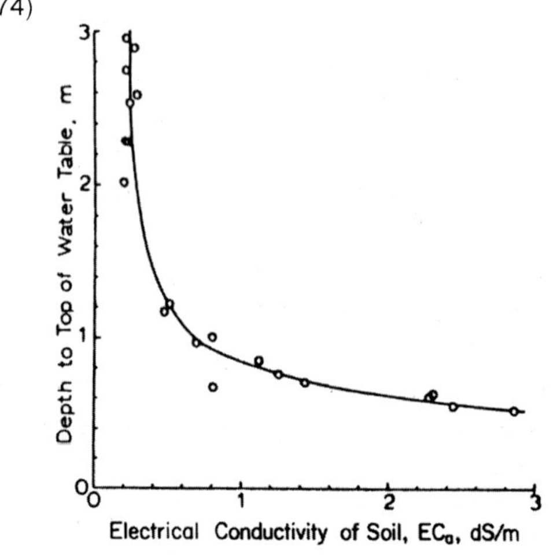

The location of the recharge area of a particular saline seep was shown to be identifiable by mapping/tracing the subsurface pattern of high EC_a levels upslope from the seep.

An example is given in Figure 55, after Halvorson and Rhoades (1976).

TABLE 2
Diagnostic EC_a values for distinguishing unaffected soil sites, incipient saline-seeps and developed saline-seeps for representative soil types of the Northern Great Plains, USA

Soil Type[1]	Site Condition		
	Unaffected	Incipient Seep	Developed Seep
	EC_a, dS/m		
C, SiC	<0.5	0.8	>2.5
SiCl, Cl, SCl, L	<0.3	0.5	>1.5
Sl	<0.2	0.4	>1.0

[1] C = clay; SiC = silty clay; SiCl = silty clay; Cl = clay loam; SCl = sandy clay loam; L = loam; Sl = sandy loam

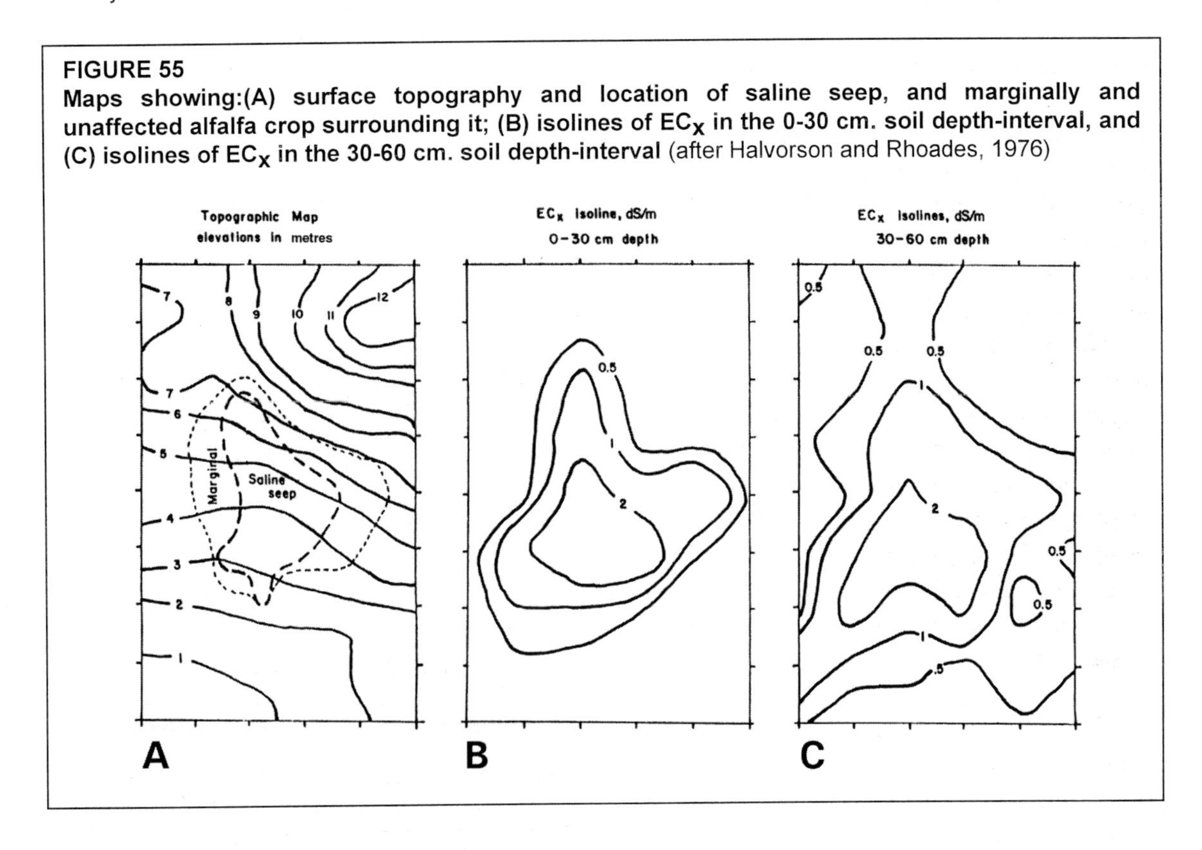

FIGURE 55
Maps showing:(A) surface topography and location of saline seep, and marginally and unaffected alfalfa crop surrounding it; (B) isolines of EC_x in the 0-30 cm. soil depth-interval, and (C) isolines of EC_x in the 30-60 cm. soil depth-interval (after Halvorson and Rhoades, 1976)

These figures show that the flow of water into the seep is from the north (there were several possibilities) and outward primarily toward the southwest. This methodology permitted the location of the recharge area to be identified and planted with permanent vegetation to help mitigate the problem.

This example illustrates how the instrumental field-salinity assessment methodology may be used to facilitate the collection of spatial information about the levels and distributions of soil salinity in dryland soils, which in turn leads to the useful determination of the eminence of a saline seep problem and to the interpretation of the sources and causes of salinization in an affected seep-area and, thus, to meaningful management implications. Of course, the measurement technology also permitted the problem of salinity to be diagnosed and the areal extent of the salinized soil in the surveyed area to be mapped. This latter topic is the subject of the following section.

While the example given in this section was based on the use of four-electrode methodology, EM methods could have also been used; however, when this work was undertaken, such equipment was not yet available. Likewise, the mobilized-sensor equipment and interpretive methodologies developed since then would also permit faster and more detailed faster surveys to be made now days.

INVENTORYING SOIL SALINITY

Both near surface and subsurface salinity have been successfully mapped using several of the instrumental approaches/methods of salinity assessment previously described by collecting the instrumental measurements in relationship to spatial location and by displaying the data in terms of maps, or as "transect" plots, of EC_a, EC_w, or EC_e.

FIGURE 56
Antenna, battery pack and meter for measuring location on the landscape using the LORAN technique (after Rhoades *et al.*, 1990c)

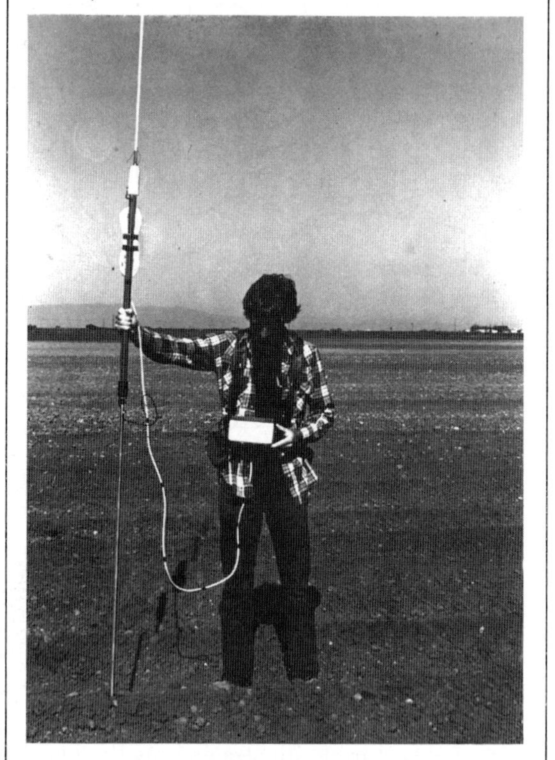

Example maps of soil salinity in the vicinity of a saline seep produced by the "soil-type" technique were shown in Figure 55. A detailed evaluation of the suitability of the various sensors for measuring EC_a and of the "EC-Model" technique (see Chapter 3, section *Procedures for measuring bulk soil electrical conductivity*) for converting EC_a to EC_e and for mapping soil salinity was undertaken in a 15-square mile (39-square km) sized study-area in California (Rhoades *et al.* 1990a). In this project, the instrumental measurements were made manually at pre-determined sites which were located by use of LORAN navigation techniques (see Figure 56; GPS equipment was not yet readily available nor affordable at the time); the area was traversed on foot (practical mobilizing equipment had not yet been developed). Contour maps were made of measured and predicted salinities using data collected at about one thousand locations. EC_e was predicted from EC_a as measured by both four-electrode (Wenner-array and insertion EC-probe) and EM-38 sensors. The values of EC_e (both measured and predicted) were plotted using SURFER (1986) software at both 200-m and 400-m grid spacings. This resulted in 1000 and 273 equally spaced grid nodes, respectively. Only contoured maps for the 400-m grid spacing are shown here (see Figures 57A and 57B), because the 200-m grid spacing resulted in too many contours to clearly represent in maps of such scale. The corresponding values of measured and predicted EC_e at each of these nodes were determined with the SURFER software. The proportions of the surveyed area by classes of soil salinity are given in Table 3. Such data can be used advantageously to assess the magnitudes of crop yield losses and associated economic losses caused by soil salinity.

FIGURE 57
Contour map of:(A) measured and (B) predicted soil salinities, in 15-square mile study area in Central California USA (after Rhoades *et al.*, 1990a)

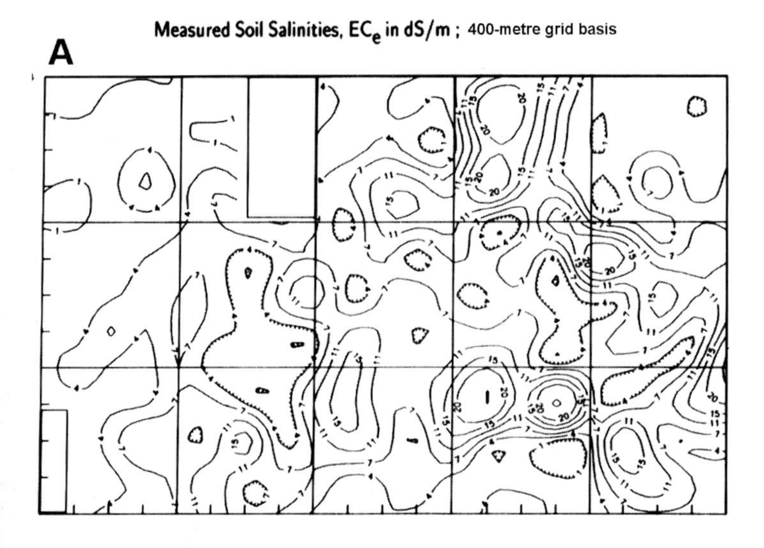

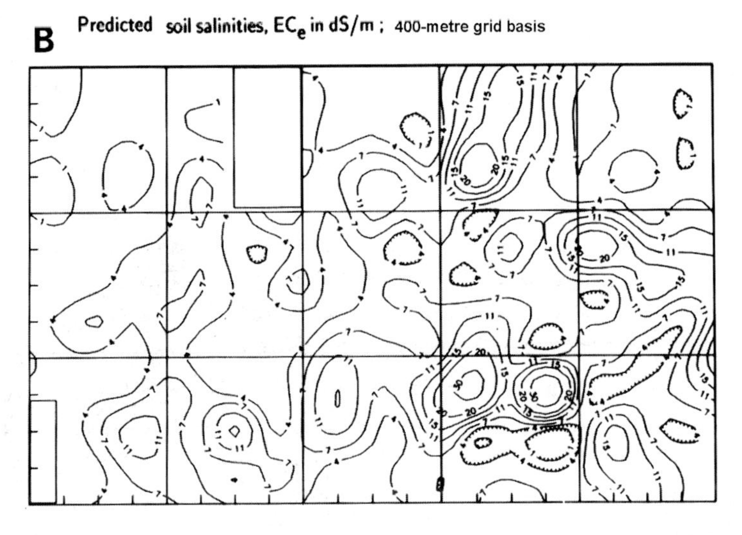

Visual comparisons of these measured and predicted salinity maps showed them to be essentially the same, irrespective of which of the three sensors was used to measure EC_a. The absolute levels of salinity estimated from the three sensors. were also similar (see Rhoades *et al.* 1990a). Where the differences were substantial, the salinity levels were so high as to make the errors in estimate agriculturally unimportant. Of the three methods used to measure EC_a , the four-electrode probe was found to be the most accurate, followed by the four-electrode surface-array and then the EM-38. The accuracy's essentially followed the degree to which the volume of soil measured by the sensor compared with that of the soil sample used to

TABLE 3
Salinity distribution in Kings River Watershed Survey Area

Quantiles	
Percentage, %	Soil Salinity; EC_e, dS/m
100	79.8 (max)
90	15.6
75	8.1 (Q3)
50	3.5 (med)
25	1.2 (Q1)
10	0.7
0	0.3 (min)

determine salinity (which was determined by conventional laboratory methods and assumed to represent "truth"). The accuracy of any of these instruments was adequate for practical salinity mapping purposes. Sample variability due to size differences in the volumes of soil used to measure salinity, which was small compared to that measured with the four-electrode surface-array and EM-38 sensors, was concluded to sometimes be appreciable and, when so, to result in an underestimate of the accuracy of the EC-model method of salinity appraisal using these larger-volume sensors under the surveyed field conditions (Rhoades *et al.* 1990a).

FIGURE 58
Maps of measurement and sampling locations and of measured and predicted soil salinity patterns in a field (Hanford S2A) located in the San Joaquin Valley of California USA (after Rhoades *et al.*, 1997d)

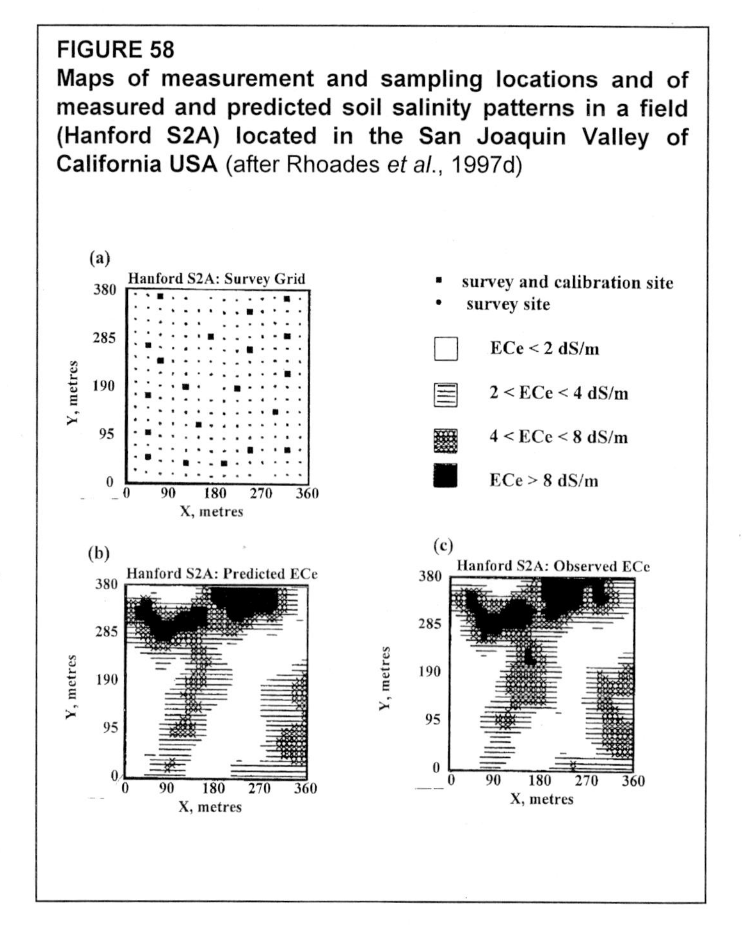

These latter findings demonstrate that soil salinity can be appraised and mapped without need for collecting any soil samples or for carrying out laboratory analyses by using sensor-measurements of soil electrical conductivity (by any of the three methods tested: four-electrode probe, four-electrode surface array, and electromagnetic induction) and simple estimates of soil water content relative to field-capacity and clay percentage made in the field by "feel" methods. The simpler, "soil-type" calibration-approach was deemed unsuitable for this situation where the soils varied so much in texture and moisture condition within individual fields and from one field to another over the area. The variability of soil-type within mapping units was concluded to be too large for this latter method to work well in this situation. However, the EC-model approach gave good results, irrespective of this variability-problem.

More recently, soil salinity has been characterized in even more detail than that described above using the mobilized four-electrode and EM systems and the newer "stochastic field-calibration" technique for converting EC_a readings to EC_e described earlier. Numerous fields have been successfully surveyed with the mobile systems, collecting readings at spacings that provided a grid-like pattern of required/desired intensity (generally between 10 and 50 m apart). The locations of the measurement and calibration-sample sites were established using the GPS technology described above. A small number of the measurement sites were selected for soil sampling (ground-truthing) based on the observed EC_a field pattern (using the ESAP software described in Chapter 3, section *Determination of locations of measurement and calibration sites*). A second trip into each surveyed field was undertaken to collect the relatively small number (usually 8-16) of soil samples using rapid, tractor-mounted augering/coring equipment. The salinities of these soil samples were then determined using the rapid method of Rhoades *et al.* (1989b). The salinities at the remaining nonsampled sites were predicted from the corresponding EC-sensor readings through use of the multiple linear regression relation (Equations [33] and [34]) established for each field using the ESAP software. The salinity contour-maps obtained using this method/software were nearly identical to those obtained by conventional soil sampling/laboratory analysis and cokriging methodology. An example of this approach is illustrated in Figure 58, after Lesch *et al.* (1992) and Rhoades (1997b). This approach provides a very practical and cost effective means to substantially reduce the number of soil samples needed to accurately map the detailed spatial salinity patterns that occur at the field scale (see Chapter 5). Such detailed spatial data can also be used to assess the adequacy/appropriateness of irrigation/drainage systems as is discussed later. This methodology is less suitable to map large areas in broader detail, such as the variable 15-square mile area discussed just above, because each field would have to be surveyed in full detail. Thus, for the less detailed broad-scale mapping purposes the "EC-model" technique is recommended; the "stochastic field-calibration" technique is recommended for detailed field-scale mapping purposes.

MONITORING SOIL SALINITY

It is important to be able to detect changes and trends that are occurring in salinity conditions and patterns over time in fields and projects, in order to be able to detect emerging problems, to evaluate the effectiveness/appropriateness of management practices, especially newly implemented ones, and to determine the progress of reclamation efforts. Traditional statistics provide tests to compare the means of two populations for differences and can be applied for some of these needs. However, the changes in the spatial levels and distributions of salinity within the soil profile and within the various fields of the project are also of interest. Formal quantitative statistics for monitoring purposes compatible with the instrumental salinity assessment technology described herein have been lacking until just very recently (Lesch *et al.*,1998). The methodology developed in the latter research utilizes the stochastic field-calibration technique of predicting salinity from EM-38 and/or four-electrode sensor measurements and a field-specific calibration based on limited soil-sample data. It is advocated herein, together with certain test-statistics for monitoring purposes, i.e., for determining if the salinity pattern of a field has changed or if the average salinity level of the entire field has changed over time. This theory/methodology was successfully tested and its utility demonstrated using the data of Diaz and Herrero (1992) and Lopez-Bruna and Herreo (1996), which is rather unique, in that it is one of the few published data sets where both EM-38 and soil salinity data were both acquired at multiple times from within the same field. Because the statistics involved in this procedure are rigorous, as well as beyond the training expected of the typical reader, they will not be presented here in detail. Rather, the interested reader is referred to the publication of Lesch *et al.*(1998) for this information and to Annex 9 where a very brief description of the test-statistics and relations are provided. However, a brief qualitative overview of the approach and tests will now be given to illustrate the general features of the monitoring approach and methodology.

The basic approach used to monitor soil salinity is: (a) first, to estimate a regression model (using Equations [33] and [34]) which is capable of predicting soil salinity at every grid site within the surveyed field from the collected sensor readings, (b) at some future time, to acquire new soil samples at two or more of the previously surveyed sites, and (c) finally, to apply the formal test-statistics described in Annex 9, in order to determine if differences exist between the reviously predicted and recently observed salinity levels and patterns. Two test-statistics are used in this comparison: (a) a test to detect a change in the fields median value of salinity between the initial and current times and (b) a test to detect any change in the dynamic spatial variation of the salinity pattern over time - i.e., to see if the salinity pattern has changed in a non-random, dynamic manner across the field. Neither of these tests require that the entire field be resurveyed nor that extensive repeated soil sampling be undertaken. For example, 15-20 soil samples are usually sufficient for monitoring purposes for establishing the initial spatial regression equation for a field; the acquisition of 8-10 new samples are typically sufficient for each subsequent period of testing. However, if a new map is desired in order to display the new pattern, presuming it has changed, a second full-survey of the field is required, as well as the development of a new regression model between the sensor readings and soil salinity appropriate for each successive testing period. Although the second set of survey data is not required to compute the test statistics, these data must still be acquired in order to create a new salinity map. Since, the changes in the pattern of salinity in a field are generally meaningful and of interest, it is recommended that the analyst repeat both the sensor readings and the calibration procedure for each subsequent time period for which significant changes in salinity condition have been detected. Whenever these surveys are to be repeated, the successive survey-sites should be co-located with the initial ones in order to permit the test-statistics to be correctly applied. One should not try to apply the predictive regression relation established between the sensor readings and soil salinity for the first survey time to any subsequent time, nor should one try to apply the test-statistics to data sets that were established on non overlapping grids.

A more general treatise on various statistical methods available to determine and map soil degradation over time is that of Hoosbeek *et al.* (1997).

DEVELOPING INFORMATION FOR SITE-SPECIFIC MANAGEMENT

Information concerning the spatial distribution of various soil physical and chemical properties within a field are needed, along with correlated plant yield relations, to optimally develop a site-specific farming plan, i.e., management that accounts for the variability of soil properties and crop differences that exist within the field. Among the soil properties of interest, besides salinity, are: infiltration rate, water-holding capacity, drainage rate, micro-relief, soil-depth, fertility, organic matter content, pH, and texture. Traditionally many of these soil properties have been estimated and mapped from laboratory analyses of soil samples collected on the basis of a relatively coarse grid, due to the lack of practical ways to measure them directly in the field. Alternatively, yield maps have been developed by spatial-samplings as a means to estimate the different crop input needs that vary among the various regions of the field. Correlations between these soil properties and plant responses are being sought to identify the causes of observed spatial-differences in yield and to develop predictive models for estimating the spatially varying farming input needs that exist within individual fields or management units. Management to compensate for field variability in salinity has not received much attention in the past, but, the author believes, it will in the future.

A limitation in the use of conventional soil sampling/laboratory analysis methodology for characterizing the spatial variability of soil properties is the high labor requirement involved. Typically for prescription farming applications, a grid of 100 by 100 metres (330 by 330 ft) is used (about one sample per ha. about one sample per 2.5 acres), which often is not intensive enough. The proper grid spacing depends upon the variability of the property of interest, which, of course, is unknown at the outset. Thus, the proper locations to collect the samples and the number of samples required cannot easily be determined by the conventional approach. As a result, too few samples are frequently taken to properly characterize the variability that often exists in fields for prescription farming purposes. No cost effective, scientific approach for determining grid size has been developed using such grid-point methods. Thus, directed sampling and remote sensing techniques are being sought and advocated, in order to site optimum soil-test locations and to minimize sampling needs. But traditional methods of directed sampling and remote sensing often do not provide enough, or sufficiently quantitative, information about the various soil properties described above for the needs of prescription farming.

On the other hand, measurements of EC_a and of geospatial position can be obtained rapidly with geophysical sensors and used to determine optimum soil-test sites as explained earlier; additionally, EC_a can be used to infer a number of soil properties, besides salinity, that are useful to prescription farming purposes and thus to create much more detailed and affordable soil-property maps than those obtained by the use of conventional soil/grid-point sampling methods (Kachanoski *et al.*, 1988; Lesch *et al.*, 1992; Doolittle *et al.*, 1994; Jayne, 1996). The theory for using measurements of soil electrical conductivity and spatial calibrations of pertinent soil properties for use in prescription farming applications is described in the paper by Rhoades *et al.* (1997d). Some example-applications for salt-affected soils follow.

FIGURE 59
Three-dimensional map of the electrical conductivity of bulk soil (EC_a) of a salt-affected field in the Coachella Valley of California USA (after Rhoades *et al.*, 1997d)

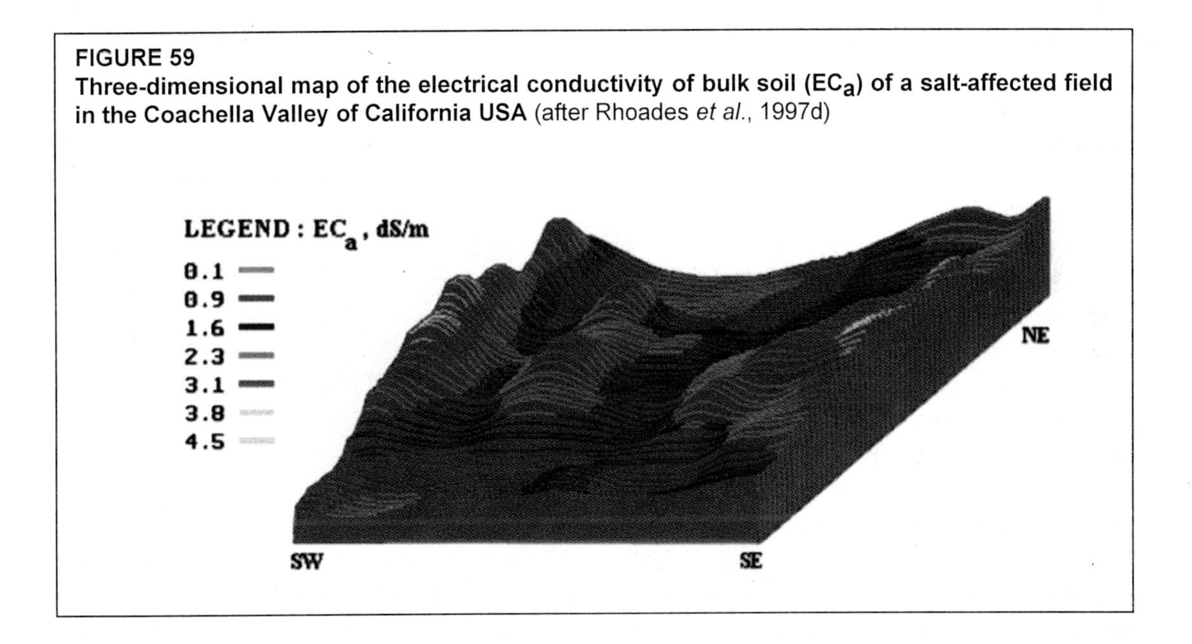

There are numerous situations where the reclamation and management needs of saline soils vary within individual fields and prescription farming methods could be used to advantage. An obvious situation is evident in Figure 59, which displays the marked variability of EC_a that existed in a salt-affected field located in the Coachella Valley of California, as measured by the mobile, combination four-electrode/EM system previously described. These data were converted to units of soil salinity by the stochastic field-calibration calibration method and corresponding maps were prepared. The results showed that salinity was exceedingly excessive (see Figure 60) for crop growth in most of the field; the median EC_e level was 43 dS/m and ranged from 3 dS/m to 106 dS/m. Additionally, soil sodicity, as expressed in terms of the sodium adsorption ratio of the saturation extract (SAR_e), was very high (see Figure 61), and quite variable, in the field (median value of 146; ranging from 9 to 475).

FIGURE 60
Estimated salinity (EC_e in dS/m) of the 0-60 cm depth-interval of a salt-affected field in the Coachella Valley of California USA (after Rhoades *et al.*, 1997d)

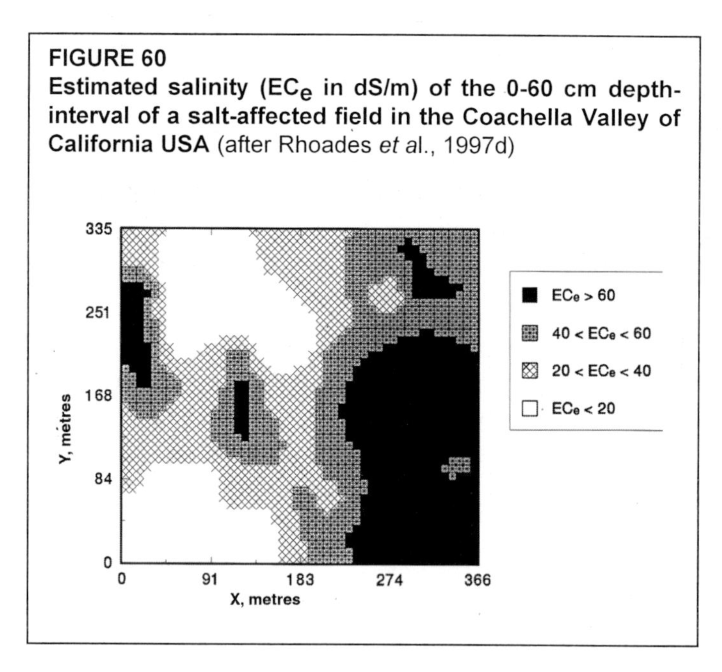

FIGURE 61
Estimated sodicity (SAR_e basis) of the 0-60 cm depth-interval of a salt-affected field in the Coachella Valley of California USA (after Rhoades *et al.*, 1997d)

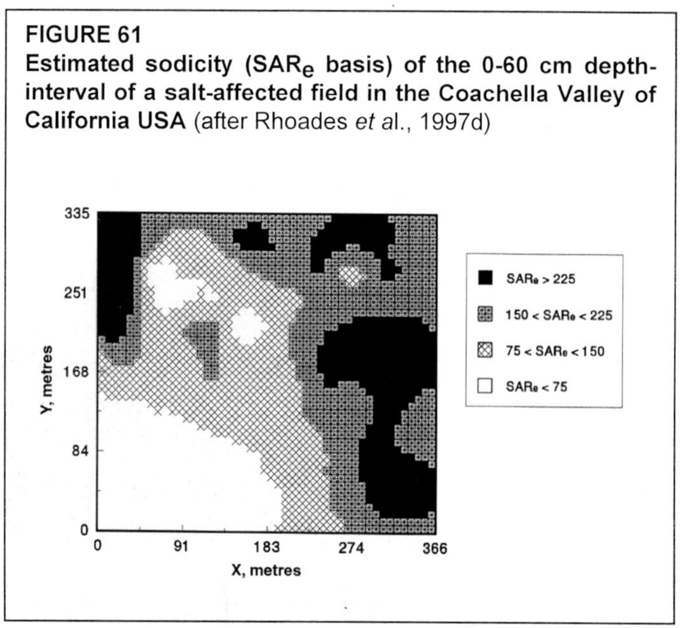

SAR_e is generally well correlated with EC_e in most fields; as is the case here, though less so than typical ($r = 0.78$). Reclamation was obviously required before this field could be successfully farmed and a plan was sought accordingly. In this regard, it was recognized that the reclamation of such saline/sodic soils may, or may not, require the use of amendments to replace exchangeable-sodium and to sustain permeability during leaching. To determine the spatial differences in the needs of gypsum for reclamation purposes in this field, the combinations of SAR_e/EC_e existing in the field, as obtained by the detailed measurements of EC_a and the analysis of soil samples collected from the "calibration" sites selected using the spatial-sampling procedures described earlier, were classified into four categories. The spatial patterns of these four conditions of SAR_e/EC_e are depicted in Figure 62. Bulk samples of soil (0-60 cm depth) were collected from these four regions, packed in permeameter-columns, with and without the addition of gypsum (2, 5 and 10 tons per acre basis; 4.5, 11.2 and 22.4 Mg per ha.) and leached with the local irrigation water. The hydraulic conductivities of these soil-columns were monitored throughout the more than five pore-volumes of leaching they were subjected to, as were the EC and pH levels of the effluents. The results obtained showed that only the regions of the field with SAR_e/EC_e ratios of greater than 5 and EC_e levels of less than 20 dS/m benefited from the addition of gypsum; without it the permeabilities of the soils in these areas decreased by more than a factor of two after 2 pore volumes of leaching, as the effluent EC decreased below 2 dS/m and its pH increased to 9.3 or greater. Given this information, it was determined that gypsum would be beneficial in only a small area of the field, that part of the field shown in Figure 62 having SAR_e/EC_e ratios greater than 5. In this manner, the reclamation prescription, in terms of gypsum requirement, was established for the different classes of chemistry and soil type existing in the different regions of this saline/sodic field.

The overall need for gypsum determined by this spatially discriminating procedure (prescription farming approach) was less than one-third of the amount that would be required to uniformly treat the field, as is traditionally done. The resulting savings in the cost of the gypsum and its application was more than US$ 25 000 for this one relatively small field (40 acres; ~ 16 ha.). The conditions of salinity and sodicity existing in salt-affected soils are typically spatially variable, not unlike those seen here. Thus, one can conclude that the reclamation requirements of typical salt-affected fields are spatially variable and can be defined and prescribed advantageously using the general approach undertaken in this example.

FIGURE 62
Estimated salinity/sodicity ratio of the 0-60 cm depth-interval of a salt-affected field in the Coachella Valley of California USA (after Rhoades *et al.*, 1997d)

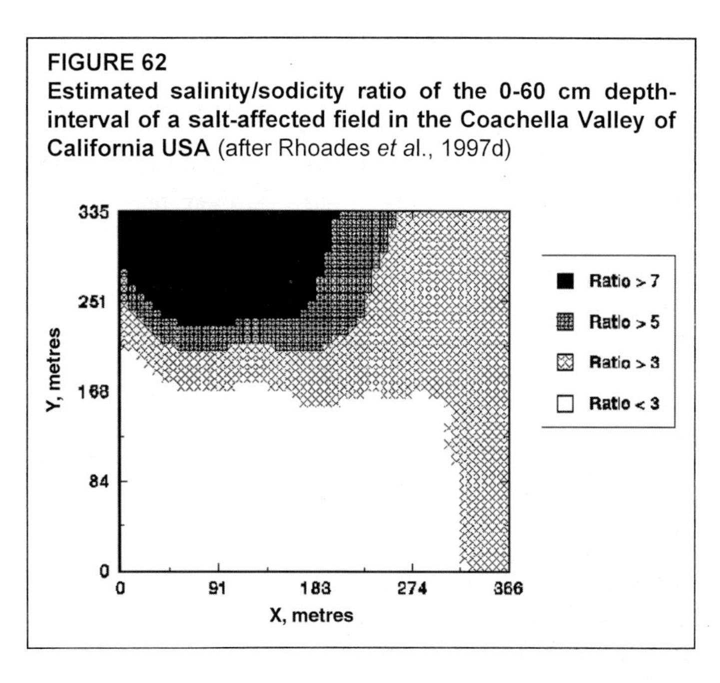

FIGURE 63
Three-dimensional map of the electrical conductivity of bulk soil (EC_a) of a salt-affected Sudangrass field in the Coachella Valley of California USA (after Rhoades *et al.*, 1997d)

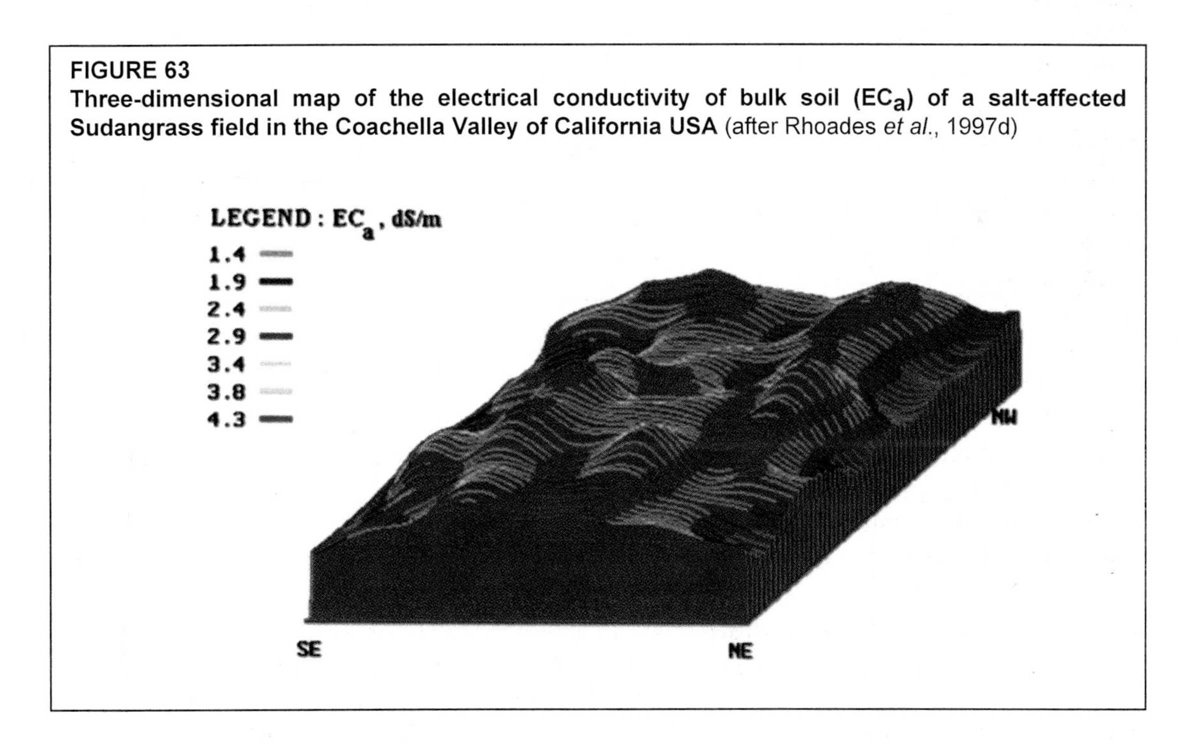

Another salt-affected field in the Coachella Valley of California that was "surveyed" with the mobilized EC_a measuring equipment is depicted in Figure 63. Like the other salt-affected field discussed above, this field also displayed substantial spatial variation in EC_a (and correspondingly in salinity and sodicity). However, in contrast to the previous field, the pattern in this second field was markedly cyclic in a north-south orientation, as illustrated in Figure 64; the "valleys" in the north-south oriented traverses correlated spatially with the presence of a sub-surface tile-drainage system (which will be discussed more later).

FIGURE 64
Relationship between the electrical conductivity of bulk soil (EC_a) and distance across (and perpendicular to the sub-surface drainage system) a salt-affected sudan-grass field in the Coachella Valley of California, USA (after Rhoades, 1996b)

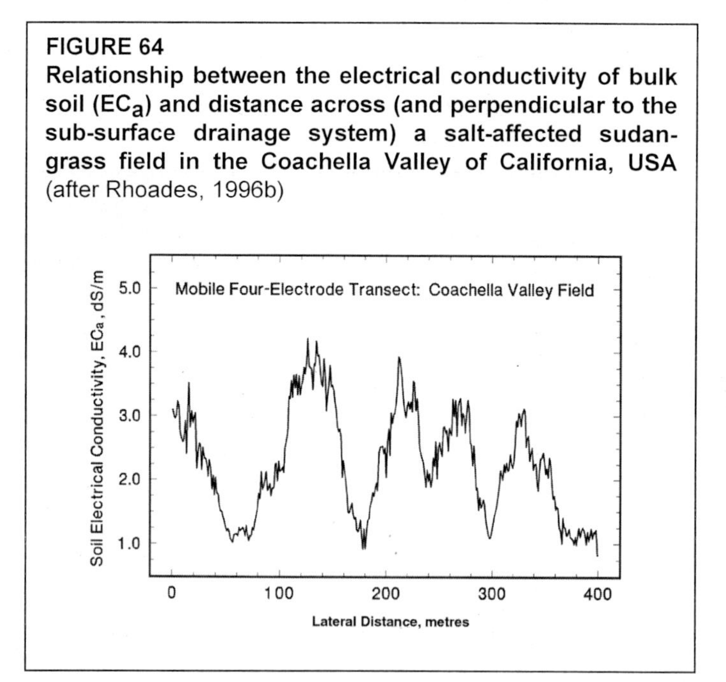

The soil salinities (EC_e basis) corresponding to the EC_a values shown varied from relatively low values of 2-3 dS/m in the regions of the field overlying the drain lines to relatively high values of 20-25 dS/m in the mid-line regions. The height of the sudan crop planted in this field was well correlated with the salinity/drainage pattern, as is illustrated in Figure 65. The spatial variation in relative sudan yield (assuming it is proportional to height) predicted from the data shown in Figures 63 and 65 is shown in Figure 66. These results lead to the conclusion that the drainage system in this field is inadequate, either because of "clogging" or insufficient capacity for the given situation of irrigation practices and local hydrology. In any case, they indicate that different parts of the field vary in their management needs. For example, the areas of high salinity have less need for fertilizer, because of the reduced crop growth there, and greater need for effective leaching and drainage. The spatial variation in input needs and management requirements existing in this field is conducive to the implementation of prescription farming methodologies. Management should be altered to increase the rate of drainage and the extent of leaching in the "midpoint" regions of the field, either through renovating the drains or decreasing their spacing. Meanwhile, the amount of fertilizer applied should be reduced in these areas and the seeding rate increased. Other management practices to reduce the level of salinity in the seedbed and to improve irrigation efficiency, as described elsewhere (Rhoades *et al.*, 1992), should also be adopted in these areas.

A third field, this one from the Imperial Valley of California, with excessive salinity in certain sections of the field, as determined by calibrated spatial measurements of EC_a, is illustrated in Figure 67. In this case, the salinity increased with distance along the "head-to-tail traverses made across the field (only one traverse is shown in Figure 67) and was excessive in the "lower-third" region of the gravity, furrow-irrigated field for the full-production of even the salt-tolerant sugarbeet crop growing there (this pattern is commonly observed in fields irrigated by such irrigation methods). The distributions of salinity observed within the root zone (data not shown) indicated that excessive water had been infiltrating the "upper" sections of the field, while inadequate amounts had been infiltrating in the "lower" sections (which will be discussed more later). Obviously this situation creates variable management needs for the differentially irrigated/salinity-affected regions of the field. Management needs to be altered to improve the uniformity of irrigation and infiltration in this field.

The latter two examples illustrate a form of prescription farming that is little mentioned and utilized, i.e., management to accommodate or mitigate the non-uniformities that occur in water application, crop-consumption, leaching and drainage in gravity-irrigated fields.

FIGURE 65
Relationship between the EM-38 sensor readings of EC_a^*, Sudangrass plant height, and distance across (and perpendicular to the sub-surface drainage system) a salt-affected Sudangrass field in the Coachella Valley of California, USA (after Rhoades *et al.*, 1997d)

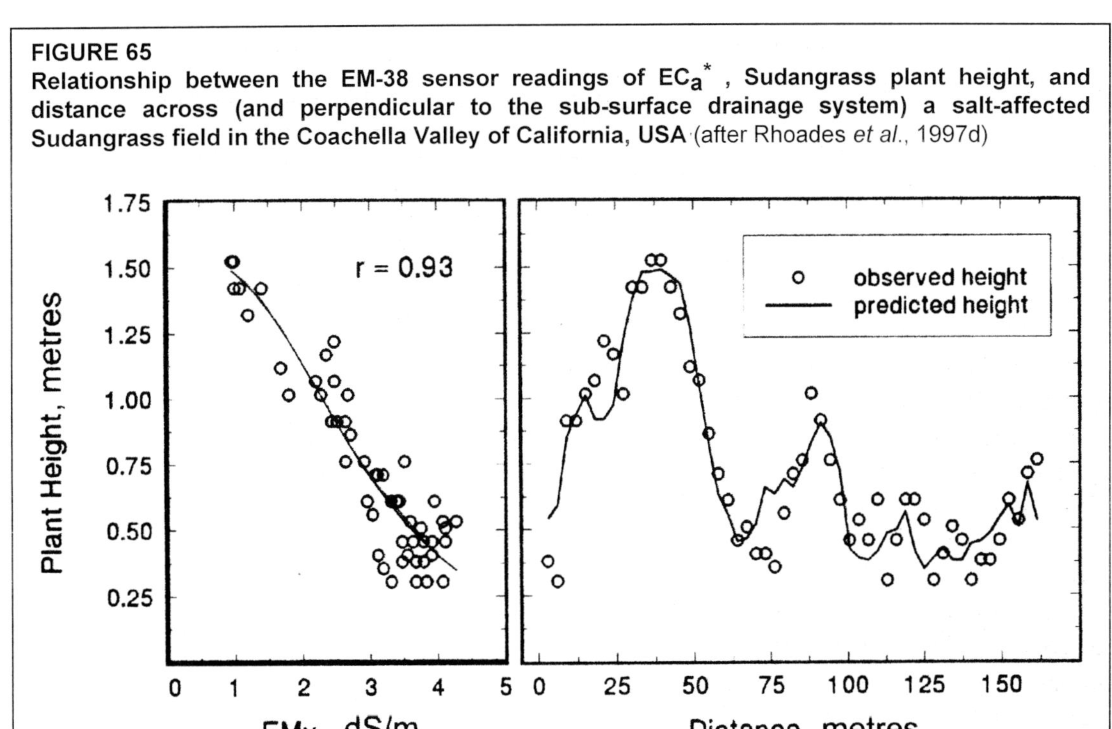

FIGURE 66
Three-dimensional map of the predicted Sudangrass yield (in terms of height) predicted from the data of Figures 64 and 65 for a salt-affected Sudangrass field in the Coachella Valley of California USA (after Rhoades *et al.*, 1997d)

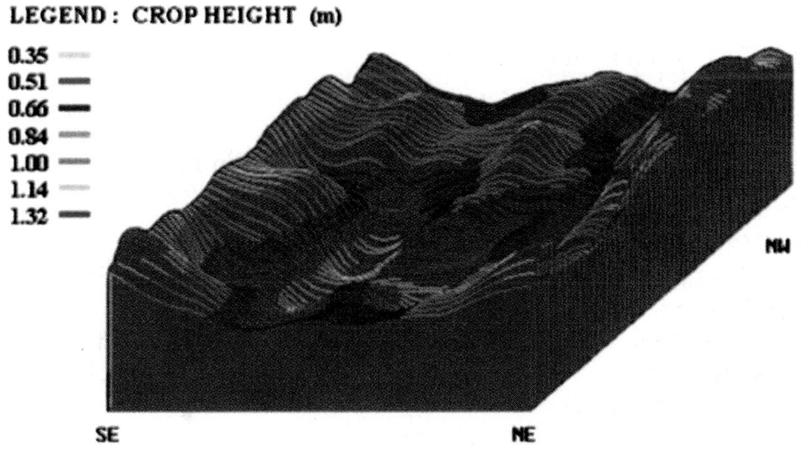

The use of the spatial measurements of EC_a and of the spatial-methods of their calibration/ interpretation, as described and advocated herein, offer a potentially valuable new tool to assess and better manage such variable-field/soil situations using prescription farming approaches. More details about the variability of salinity in the fields used in these two examples, and in others, are given elsewhere (Rhoades, 1992b, 1994, 1996b; Rhoades *et al.*, 1997a, 1997b). More details about the use of spatial measurements of soil salinity and electrical conductivity for prescription farming purposes is described in Rhoades *et al.* (1997d).

FIGURE 67
Relationship between (a) soil electrical conductivity (EC_a), as measured by both the mobilized, four-electrode system and the mobilized, electromagnetic (EM) system, and (b) measured and predicted soil salinity (EC_e basis) and distance along a transect across a furrow-irrigated, sugar beet field located in the Imperial Valley of California, USA (after Rhoades, 1997)

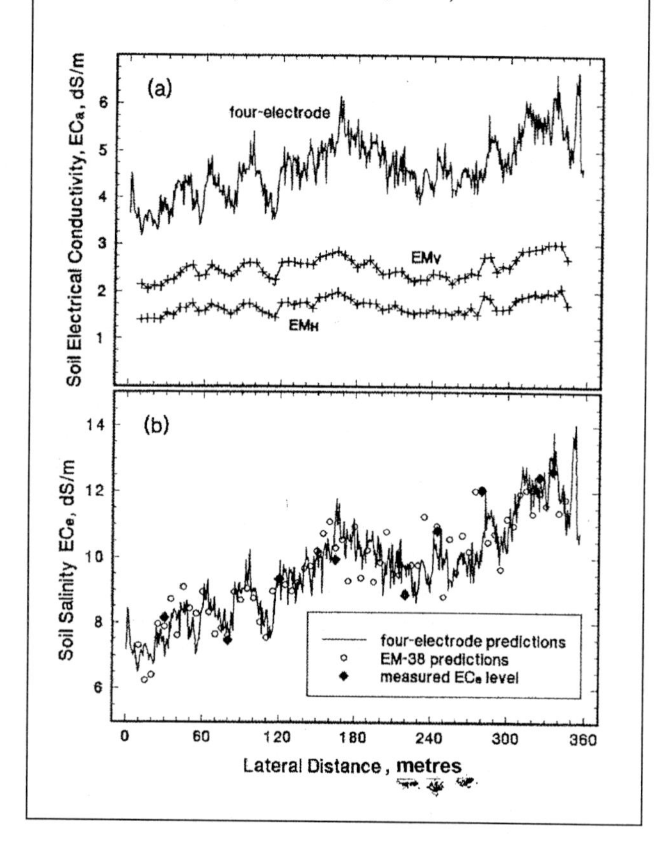

EVALUATING ADEQUACY AND APPROPRIATENESS OF IRRIGATION/DRAINAGE

As mentioned earlier, the distribution of salinity within the root zone of a soil is a reflection of the direction of the past net flux of water flow. A net upward flow, such as may occur in the presence of a shallow water table or otherwise poorly drained situation, is reflected by the presence of high salinity in the near-surface depth of soil and by decreasing levels with depth to a minimum level determined by the salinity of the shallow groundwater. On the other hand, a net and excessive downward flux of water through the soil is reflected by low levels (controlled by the salinity of the irrigation water) of salinity in the near-surface depth of soil with relatively little increase in the deeper depths. Evidence of the credibility of this generalization was presented earlier for the saline seep situation. Other examples for the case of irrigated soils will now be given to further demonstrate the utility of the salinity assessment technology to evaluate the adequacy/appropriateness of irrigation and drainage systems and practices.

The data obtained with the mobile, four-electrode and EM sensing systems presented in Figure 67 can be used to demonstrate this utility. Average rootzone soil salinities, expressed in terms of EC_e, as predicted from measured EC_a data (Figure 67a) obtained in a furrow irrigated, sugar beet field (Glenbar silty clay loam soil) in the Imperial Valley of California and as measured in some "calibration" samples collected along the transect are shown in Figure 67b. The salinity values were predicted from the sensor readings and limited calibration information, using the "stochastic field-calibration" technique. As is shown here, the accuracy of these predictions is very good. If irrigation application and infiltration were uniform across the field involved with Figure 67, the value of EC_a (and of EC_e) should be the same at each distance, provided crop stand and soil type were also uniform. However in this case, the EC_a (and EC_e) values increased from the "head" to the "tail end" of the field; the coefficient of variability (CV) was 14.2% and the linear correlation coefficient (r) between EC_a and distance down the transect was 0.67. Thus, one may conclude from these data that the field is not uniform with respect to one or more of the three possibilities. In this case, the crop was planted uniformly and the soil type was essentially the same along the transect. Hence, these observations/data imply that irrigation application, or infiltration, was not uniform across this field, presumably due to reduced opportunity-time and infiltration of irrigation water with distance from the point of water delivery to the furrows. Another factor likely influencing the salinity distribution within this field is the lateral transport of salt that occurred in it as a consequence of the "cracking" type of soil present in the field. This latter aspect is discussed elsewhere (Rhoades *et al.*, 1997b). This example illustrates how information about the spatial variation of average root zone soil salinity can be used, assuming it is a tracer of the interactions of water infiltration, evapotranspiration, leaching and drainage, to evaluate irrigation uniformity in fields which are relatively uniform in soil type and cropping intensity.

FIGURE 68
Correspondence between soil salinity predictions based on soil electrical conductivity measurements obtained with the mobile salinity assessment system along a transect across a surface irrigated, tile-drained alfalfa field (Imperial clay soil) located in the Imperial Valley of California (after Rhoades, 1996b)

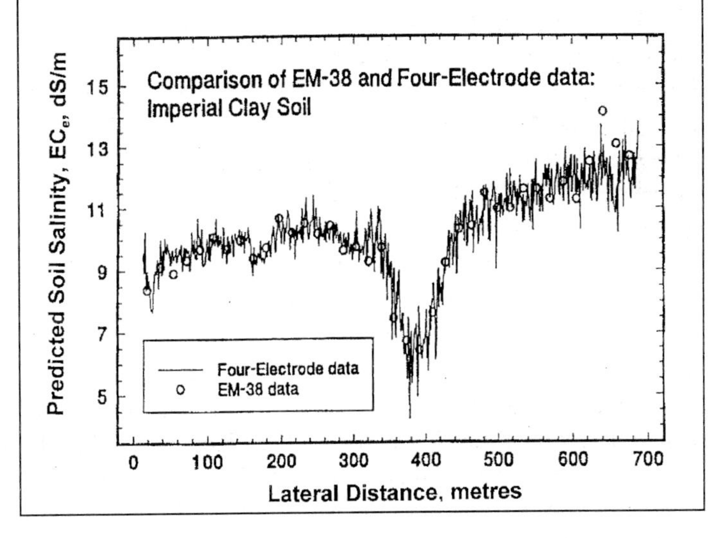

Another example of spatial-data obtained in an irrigated/drained field with the mobile, four-electrode sensing system to a depth of 1.2 metres is presented in Figure 68. This figure shows EC_e values calculated using the stochastic field-calibration method from EC_a readings collected every second (about every 1 m apart) as the tractor moved across a furrow irrigated, tile-drained alfalfa field (Imperial clay soil) in the Imperial Valley of California. The "minimum" in the EC_e readings occurring at about 380 metres from the irrigation-intake end of the field corresponds to the position of a suite of subsurface drains. Otherwise, the EC_e values increased toward the "tail end" of the field, presumably due to reduced application and infiltration of irrigation water with distance from the point of water delivery to the furrows and to lateral transport of salt across the field as a consequence of the soil cracking and lateral-transport phenomena previously mentioned (Rhoades *et al.*, 1997b). Examples of fields with much greater increases in "tail end" salinity have been observed in other fields (Rhoades, 1992b). Tile drains are also located in the tail end of the field involved in Figure 68, but were ineffective in lowering the salinity except in the four narrow regions, indicated by the sharp downward spikes apparent in the "curve", where the trenching had occurred many years (decades) earlier. These data suggest that much of the variability in average root zone salinity observed in typical irrigated/drained fields is caused by the interactive, effects of the drainage and irrigation systems and trenching operations. They also demonstrate how the adequacy or inadequacy of a drainage system can be inferred from the detailed spatial salinity information provided by the mobilized salinity assessment systems.

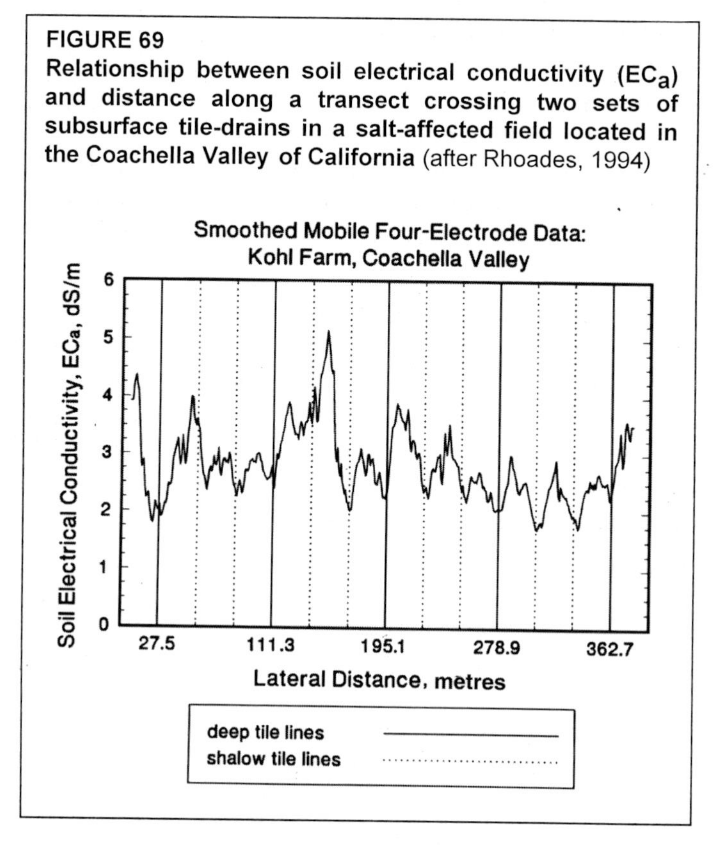

FIGURE 69
Relationship between soil electrical conductivity (EC_a) and distance along a transect crossing two sets of subsurface tile-drains in a salt-affected field located in the Coachella Valley of California (after Rhoades, 1994)

Another example of the marked effect that a subsurface drainage system can have on average root zone salinity is given in Figure 69 in terms of EC_a. Corresponding values of EC_e (not shown) cycled between low values of about 2.5 dS/m to high values of about 25 dS/m. The CV and r values for this EC_a- distance traverse were 36.8 % and -0.20 respectively. This example involves a field of silty loam soil in the Coachella Valley of California which has two sets of buried "tile-lines" oriented perpendicular to the direction of the EC_a traverse; one set being about 2.7 m deep and spaced about 90 m apart and another set being about 1.7 m deep and located at one-third and two-third distances between the deeper lines. The two sets of tile-lines are represented by the solid and dashed lines, respectively, in the figure. In this field, soil salinity levels “mimicked” the drainage system, with high values of EC_a (and EC_e) occurring in the soil located between tile-spacings and low values in the soil overlying them. These data suggest that most of the variability in average root zone salinity across this field was caused by the effects of the drainage system. They also imply that the drainage system there was inadequate given the circumstances of irrigation, soil type, geohydrology, etc. The distributions of salinity within the root zone depth of such fields will be discussed later; they give further credence to the preceding conclusion.

The time involved in collecting these data was only about seven minutes. Again, they demonstrate the utility of the assessment methods for evaluating the adequacy of the drainage conditions of the field. An even more dramatic drainage system effect on soil salinity is evident in the previously discussed Figure 64; again, such data show the utility of the rapid salinity assessment equipment and methodology.

The spatial pattern (average root zone basis) of the field depicted in Figure 69 is shown in Figure 70. The average profile EC_e value of 10-12 dS/m measured within the 0-1.2 m depth in much of this field is excessive for successful crop production. This observation itself is evidence of the inadequacy of the past irrigation and drainage management in the field. Assuming uniform irrigation and a leaching fraction of 0.05, the expected value of average root zone salinity (as calculated using WATSUIT, Rhoades *et al.*,1992) would be about 2.1 dS/m under steady-state conditions of irrigation with the Colorado River water. Since the average soil-profile salinity in this field of silty-loam soil (non-cracking soil) exceeds 2.1 dS/m, one must conclude that the overall leaching fraction is negative either because of deficit-irrigation or because salt is being accumulated in the root zone from the upward flow of saline water from the shallow groundwater. Since information supplied by the irrigator showed that the applied water exceeded ET, the latter cause is deduced. The salinity distributions found within the profiles over much of this field are presented in Figure 71; they are concluded to be affected by the drainage system. As discussed above, lower salinities occurred in this field in the soil overlying the tile-lines and higher salinities occurred in the soil located in between the tile lines. Additionally in this field, as shown in Figure 71, the distribution of salinity in the soil profile varied with the mean level of salinity. These distributions imply that salinity is high in areas where the net flux of water has been upward in the field (in the region of the field located in between the drain lines) and is low in the areas where the flux has been downward, i.e., where leaching has occurred in the soil overlying the tile lines. Figure 72 portrays the salinity distribution in the upper part of the root zone (0-0.5 m) of the Coachella Valley field. These data indicate that the salinity levels and patterns within the seed bed of this field are also related to the mean profile salinity levels, which in turn are related to the drainage pattern. Taken together, all these data (Figures 69 to 72) indicate that the drainage system in this Coachella Valley field is inadequate given the manner of irrigation, or geohydrologic situation, or both, existing there. As shown in Figure 72, the salinity distributions in this Coachella Valley silty-loam soil are clearly two-dimensional in contrast to the one-dimensional profiles observed for the clay textured Imperial Valley soil (see Figure 73, after Rhoades *et al.*, 1977b). This latter figure shows that salinity in the center of the seed-bed of the fine-textured soil is not as high as might be expected. A likely reason for this is the presence of an extensive network of cracks within the bed which allowed water movement through it, especially in the later stages of the irrigation season (Rhoades *et al.*, 1997b).

FIGURE 70
Predicted average root zone soil salinity in a tile-drained field located in the Coachella Valley of California, USA (after Rhoades, 1997a)

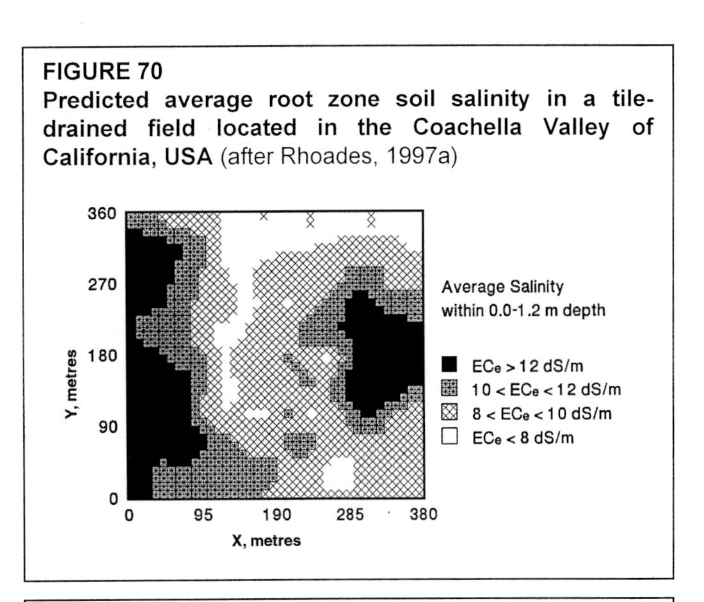

FIGURE 71
Relationship between salinity distribution and mean level of salinity in the soil profile of a tile-drained field located in the Coachella Valley of California, USA (after Rhoades *et al.*, 1997a)

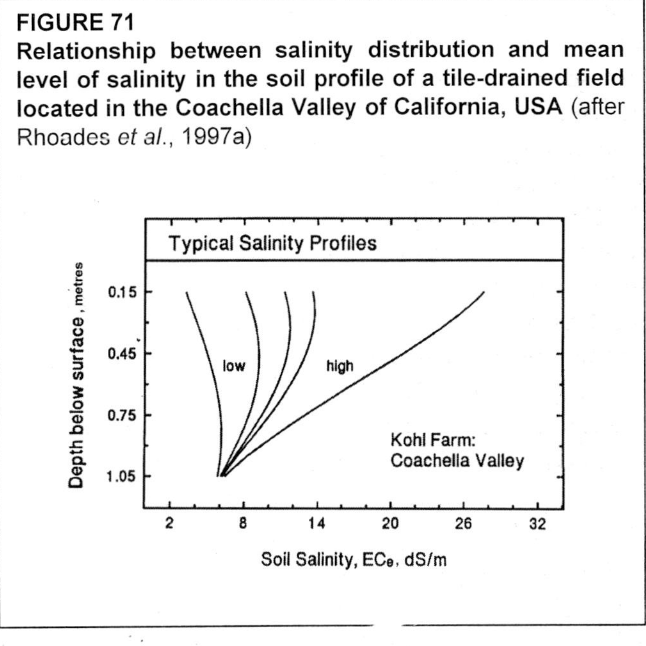

This "inter-flow" likely leached out salts which otherwise would have accumulated by capillarity and upward flow and evaporation of water in the bed, if it was completely isolated

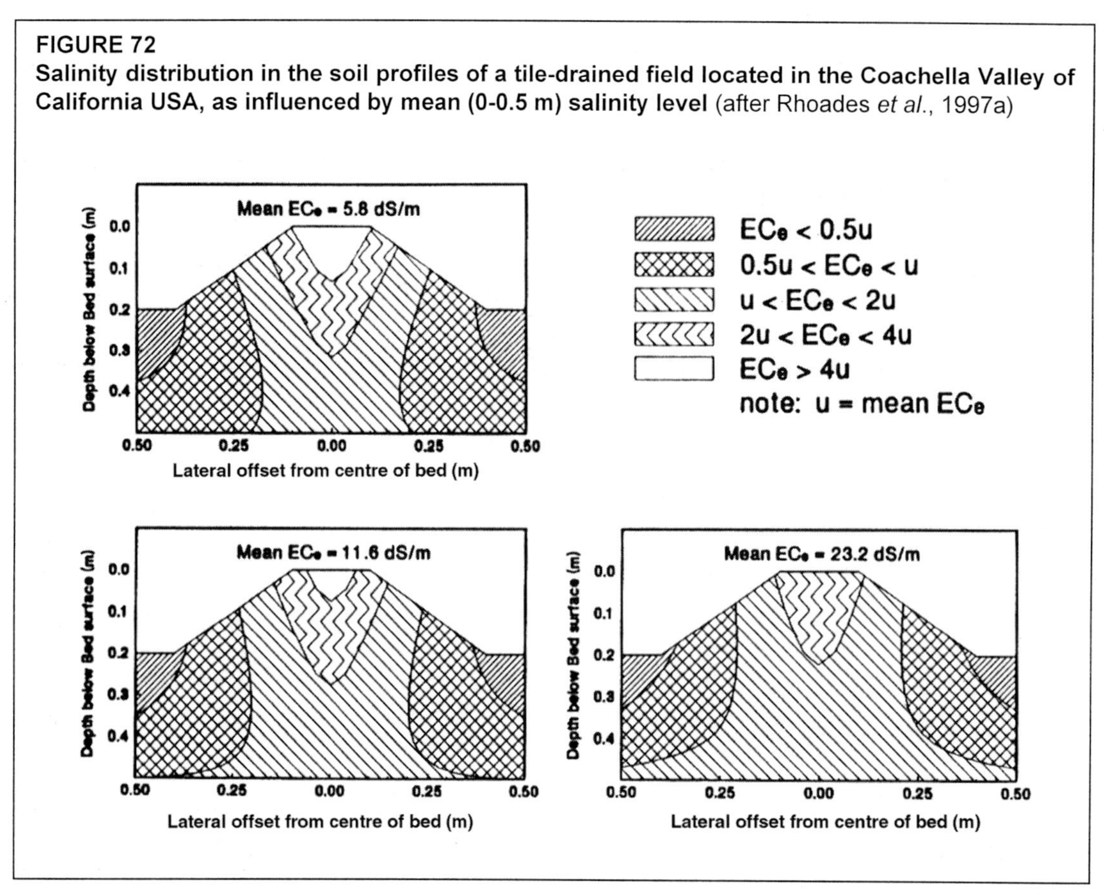

FIGURE 72
Salinity distribution in the soil profiles of a tile-drained field located in the Coachella Valley of California USA, as influenced by mean (0-0.5 m) salinity level (after Rhoades *et al.*, 1997a)

from the furrows. The patterns of salinity within the soil profiles of the Imperial Valley soil were very similar at various points along the transect; however, in relation to the average profile shape, salinity increased in the upper part of the profile and decreased in the lower part of the profile with distance towards the down gradient end of the furrow-irrigated field (see Figure 7 in Rhoades *et al.*, 1997b).

Salinity "distribution" data obtained with the "combination sensor system" in two other fields (both near each other in the San Joaquin Valley of California) are given in Tables 4 and 5 to further illustrate how information about the levels and distributions of salinity within the rootzone obtained with the mobilized EC-sensor systems can be used advantageously to evaluate the adequacies of salinity control and irrigation and drainage management. The percentages of the Borba-farm field having levels of salinities with

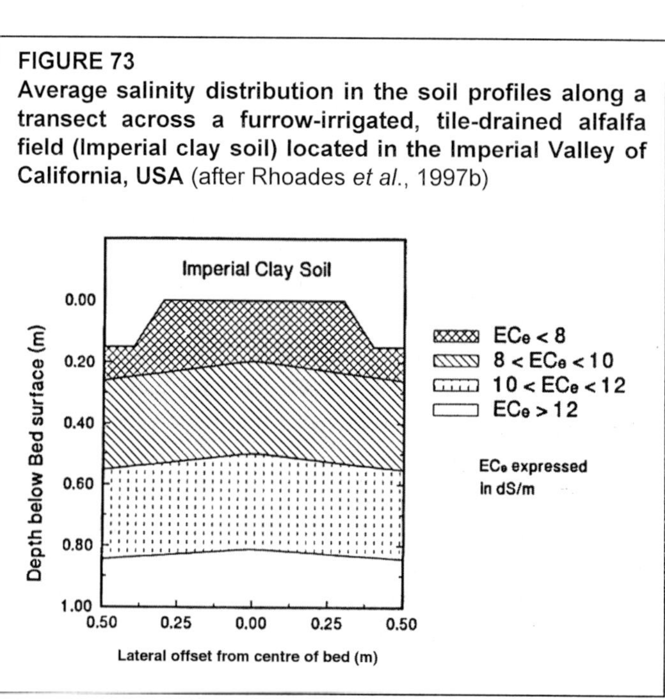

FIGURE 73
Average salinity distribution in the soil profiles along a transect across a furrow-irrigated, tile-drained alfalfa field (Imperial clay soil) located in the Imperial Valley of California, USA (after Rhoades *et al.*, 1997b)

certain ranges are given in Table 4. By reference to salt-tolerant tables, one can use these data to estimate how much yield loss caused by such salinity conditions would result for any given crop. For example, assuming the crop is alfalfa (which has a threshold EC_e value of 2.0 dS/m and a rate of yield loss of 13% for each unit of EC_e in excess of 2.0; Maas, 1990) and its effective depth of rooting is 1.2 metres, one would estimate the relative yield loss due to salinity to be as follows by percentages of the Borba field: 0% loss in 3% of the field, 14.6% loss in 49% of the field, 44% loss in 29% of the field, and 100% loss in 18% of the field. Thus, on a whole field basis, the expected salinity induced loss in relative alfalfa yield would be 38%. The economic significance of this yield loss in turn can be calculated given other cost information and used to evaluate the economic impact of salinity on the profit-line of the operation of this field and also to evaluate the affordability of improving the management to eliminate the salinity-induced yield losses.

TABLE 4
Percentages of field with soil salinities (EC_e) within certain ranges

Soil salinity	Soil depth, metres				
dS/m	0-0.3	0.3-0.6	0.6-0.9	0.9-1.2	0-1.2
0-2	14	44	17	15	3
2-4	41	32	34	31	49
4-8	36	17	22	25	29
8-16	9	6	16	17	16
<16	0	1	10	11	2

TABLE 5
Percentages of fields by different soil salinity (EC_e basis) profile types

Profile characterization		% area	
Salinity Profile Ratio	Profile Type	Furrow-field	Sprinkler-field
> 0.75	very negative leaching	5	0
0.50-0.75	negative leaching	23	3
0.35-0.50	excessive leaching	17	13
0.20-0.35	normal leaching	35	71
< 0.20	low leaching	20	13

As explained earlier, the information of salinity by depth and location in the soil profiles of irrigated fields, as is provided by the mobilized EC-38 sensor system, can be used to assess the adequacy of the past leaching and drainage practices. For example, where salinity is high in the near-surface soil of non-deficit irrigated fields and decreases with depth in the profile, the net flux of water (and salt) can be inferred as having been upward. This is reflective of inadequate drainage. Where salinity increases with depth in the profile, the net flux of water and salt can be inferred as having been downward. When salinity is low and relatively uniform with depth, leaching can be inferred as having been excessive, probably contributing to water-logging elsewhere. As shown previously (Table 29 in Rhoades *et al.*, 1992), an approximate relationship can be established between steady-state leaching fraction (L) and the ratio: EC_e in the top-half of the root zone/the sum of EC_e throughout the profile. This relationship (see Figure 74) between L and the latter ratio (salinity profile ratio, P) is: $L = 0.01843\ (e^{8.0P})$. Thus, one can infer the approximate degree of leaching (under steady-state conditions) from the salinity profile ratio, which, in turn, can be determined from the data acquired with the mobilized EM-38 sensor system. As an example, the percentages of a furrow-irrigated cotton field in the San Joaquin Valley of California are given in Table 5 by classes of salinity profile ratios. Inverted salinity profiles ($P > 0.50$) occurred in 28% of this field. Such profiles are indicative of the net upward flux of water for the reasons previously given. The author speculates, knowing that the irrigator applied water in excess of ET in this field, that excessive deep percolation occurred in the pre-season and early-season irrigations, causing a "mounded, perched" water table which was the source of the water and salt that subsequently "subbed" back up into the root zone. Profiles with salinity distributions indicative of excessive net-leaching (L values of greater than 0.3; salinity profile ratios of 0.35-0.50) occurred in 17% of the field, and profiles with salinity distributions indicative of normal leaching (L values of less than 0.3; salinity profile ratios < 0.35; salinity increasing with depth) occurred in only 55% of the field. Such data indicate that the leaching/drainage management this field has received was inadequate over much of the field. The analogous percentages obtained in a nearby San Joaquin Valley field (this one sprinkler irrigated) are also given in Table 5. While both fields were of the same soil type (SiCl) and water table depth (~1.5 m), quite different results were obtained. Hardly any of the sprinkler-irrigated field had inverted (net upward-flux; P>0.5) profiles; the desired normal leaching profiles were evident over 84% of the field.

These examples show that improved irrigation, drainage and salinity management that can result from the use of the more efficient and uniform method of sprinkler irrigation compared to furrow-irrigation. These data further illustrate the utility of the mobilized assessment system and of detailed spatial information of soil salinity and its distribution through the root zone to evaluate the adequacy and effectiveness of irrigation and drainage systems and practices. Maps of the areas with excessive leaching or of inadequate drainage can easily be prepared

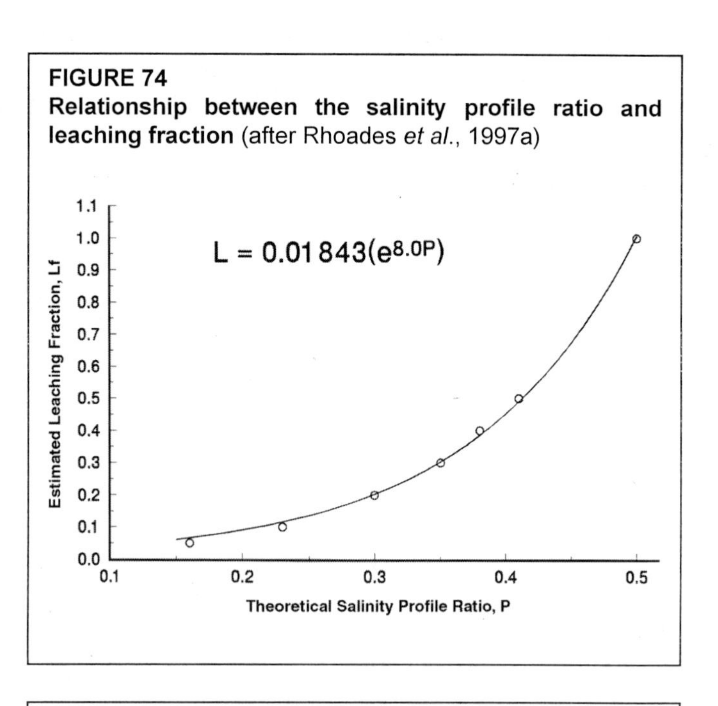

FIGURE 74
Relationship between the salinity profile ratio and leaching fraction (after Rhoades *et al.*, 1997a)

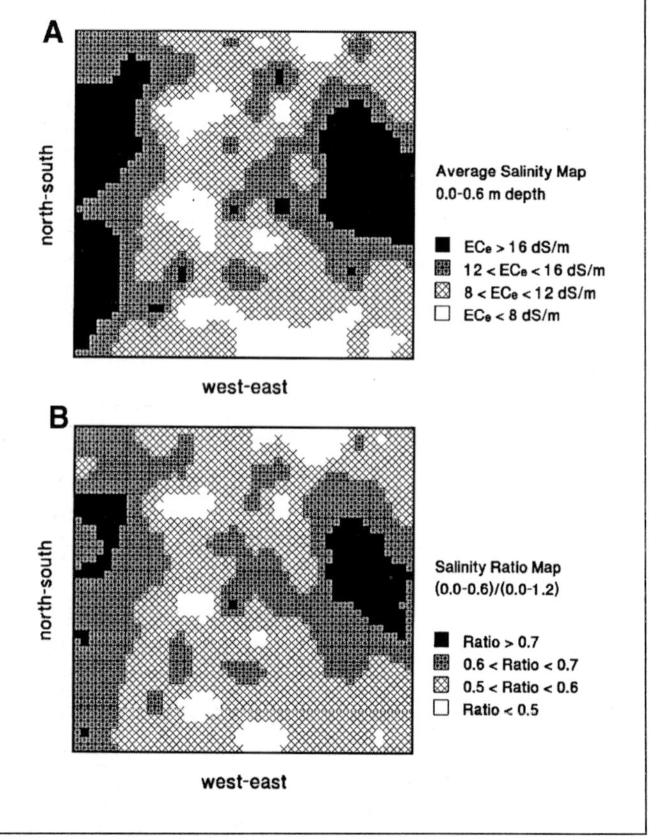

FIGURE 75
(A) Predicted soil salinity (EC_e) and (B) salinity profile ratio for a tile-drained field located in the Coachella Valley of California, USA (after Rhoades, 1997)

from these data to display the areal extent and locations of these conditions. An example of such an application is given in Figure 75, which shows the pattern of average soil salinity in the 0 to 60 cm. depth (Figure 75a) of an irrigated field (Kohl-5) located in the Coachella Valley of California and the corresponding pattern of the salinity profile-shape ratio (Figure 75b) predicted from the calibrated sensor readings. Profile-ratio values exceeding 0.5 imply inadequate drainage (a net upward flow of water) for steady-state conditions (the interpretation of the profile-ratio is discussed more later). These data imply that salinity in the major rootzone depth (0-60 cm.) is high in the areas of the field which are the least well drained, in fact in regions where the net flux of water through the root zone over the past years of cropping has been upward.

Besides irrigation and drainage, tillage and tractor traffic-patterns have also been deduced from the intensive, spatially referenced data sets obtained using the mobilized EC-sensor systems to significantly affect soil salinity levels and distributions in some fields. Tractors typically move through the fields in a systematic way, as dictated by the invoked practices of seed-bed/furrow preparation, cultivation and tillage. As a result, tractor weight is repeatedly exerted in some furrows, but not in others, leading to cyclic patterns of compaction among some sequential sets of neighboring furrows. Similarly, tillage and cultivation operations are often implemented using equipment with guide/depth wheels which similarly lead to other analogous definable "traffic" patterns. As a result, some furrows can become more compacted than others leading to reduced water-intake rates and to relatively increased lateral water flow rates and, hence, to higher salinity levels in both the associated furrows and beds. Systematic, cyclic differences have been observed in the salinity patterns of some irrigated fields surveyed with the mobilized EC-sensor equipment and to "mimic" the traffic patterns undertaken with the tillage equipment. An example is shown in Figure 76, in which the EC_a readings obtained in a succession of neighbouring furrows are presented (Figure 76a). The furrows in which the tractor tires traveled are indicated by a small inverted triangle. The EC_a values associated with the "spline-fit" (the plot of the "running average" of neighbouring values) of the readings are indicated by the dotted line. The differences between the individual EC_a values for each furrow and its spline-fitted value are presented in Figure 76b. These data show that EC_a (and, by implication, salinity) is substantially higher in each furrow in which the tractor tires traveled compared to its neighboring furrows. They also show that EC_a (thus salinity) is substantially lower in each

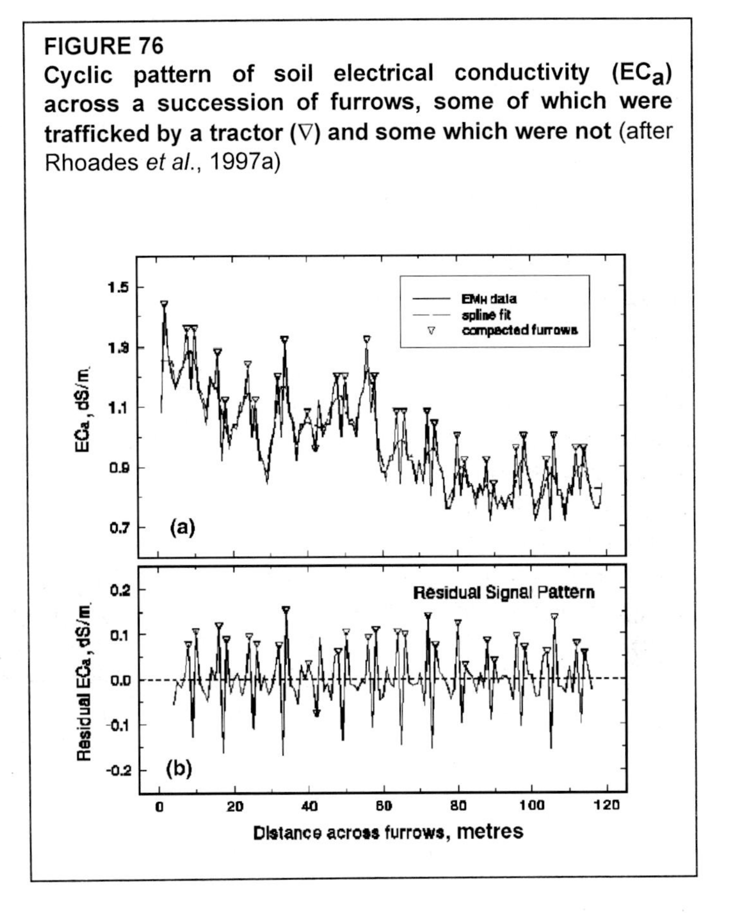

FIGURE 76
Cyclic pattern of soil electrical conductivity (EC_a) across a succession of furrows, some of which were trafficked by a tractor (∇) and some which were not (after Rhoades *et al.*, 1997a)

furrow that is "sandwiched" between "traveled" furrows. The other furrows have EC_a values that are only slightly higher, or lower, than its neighbors, as would be expected if there was no cyclic pattern or significant difference between them (that is, if all the furrows were essentially the same in their degree of compaction). The observed salinity pattern across this succession of furrows was clearly cyclic in nature and related to the tractor traffic pattern that had been followed in the field. In some fields which displayed this same phenomenon, the EC_e values in adjacent beds of furrow-irrigated fields have differed from their neighbors by as much as 4 dS/m, or more. Analogous cyclic patterns of soil salinity have been observed in other "surveyed" fields that were caused by deep chiseling actions of subsurface tillage operations. In this case, the data obtained led to the inference that water had infiltrated and flowed preferentially in the tillage "slits", then flowed horizontally out into the adjacent soil causing salinity to be lower in the vicinity of the "slit" compared to the inter-slit soil areas (data not shown). In one "surveyed" field which had been "ripped" to 0.5 m with chisels, markedly abrupt cyclic patterns of EC_a were observed that mimicked the tillage pattern. An excavation and detailed examination of the soil profile was made at the cyclic locations where the abrupt changes in EC_a were measured. This examination revealed (once the topsoil was removed) the presence of deep narrow trenches, or cracks, approximately 2.5 cm wide in the soil underlying the topsoil mulch. An interesting feature of these "cracks" was that they were full of dry aggregates of **surface soil** that had fallen down into them. Hence, such "cracks" not only provide preferential paths for water flow, but as well provide a means for soil particles and associated organic matter to "fall" to deeper depths in the soil profile and thus a means by which certain pesticides and other relatively immobile chemicals may be transported in soils that is not accounted for in classical solute transport theory. This observation would not have been made without the use of the detailed spatial measurement system.

The examples given above indicate how the salinity patterns within fields and root zones can be used as tracers of the net interactions of irrigation/infiltration, evapotranspiration and drainage to deduce much useful information about the adequacy, uniformity and appropriateness of the irrigation/drainage/salinity management. The required spatial data can be practically acquired using the mobilized systems of salinity assessment. Ways to obtain more quantitative interpretations of the amounts of leaching and its associated salt-loads are discussed in the next section.

ASSESSING LEACHING AND SALT-LOADING

While salt–affected soils and waters occur extensively under natural conditions, the salt problems of greatest importance to agriculture arise when previously productive soil and water resources become salinized as a result of agricultural activities (so-called secondary salinization). The extent of salt-affected soil and water resources has been influenced considerably by the redistribution of water (hence salt) through irrigation and drainage.

Essentially the same processes are the root causes of both soil and water salinization (Rhoades, 1997a). Salinization comes about primarily as a result of the processes described in the following scenario. Water containing salt is applied in excess of that which the crop can use and the soil can retain, in at least in some of the irrigations and/or in some parts of the field. The excess water passes beyond the root zone as drainage flow containing most of the applied salts in a reduced volume and proportionately increased concentration. This water, together with that percolating downward from canal seepage, dissolves additional salts (over and above those present in the applied water) from the soil and underlying substrata. Such concentrated and additionally mobilized salts, when transported to lower-lying landscapes and receiving waters, result in excessive salt accumulations in these areas, i.e., in soil degradation and/or in water pollution. From the preceding it is clear that the major source-areas of salt-loading in irrigated lands are the regions where the application of irrigation water is high, the leaching fraction is high and the substrata contains readily dissolvable salts within them. In order to determine the salt-loading coming from the root zone, one needs to be able to measure, or calculate, the volume and concentration of deep percolation. As mentioned earlier, the protection of our soil and water resources against excessive salinization, while sustaining agricultural production through irrigation, requires the ability to determine the areas in irrigation projects and in individual fields where excessive deep percolation is occurring, i.e., where the water- and salt-loading contributions to the underlying groundwater and surface water are coming from (a suitable means of determining areal sources of pollution). Though less critical, it is also useful to know the amounts of leaching and associated salt-loading.

As explained earlier, the level and distribution of salinity in an irrigated soil is a reflection of the net interactions of irrigation, evapotranspiration, leaching and drainage and, thus, may be used as a tracer in this regard. The relationships that have been developed between soil salinity, leaching and salt-loading, with reference to the soil profile, will be discussed in this Section. These relationships have been based primarily on simple salt- and water-balance concepts, though some refinements can and have been attempted to adjust them as needed to account for certain deterministic processes. The following equation describes the major inputs and outputs of salts that are involved in the salt-balance of the rootzones of cropped soils:

$$salt\ input = salt\ output \pm \Delta\, soil\ salinity\,, \qquad [35a]$$

$$V_{iw}C_{iw} + V_{gw}C_{gw} + S_m + S_f = V_{dw}C_{dw} + V_{tw}C_{tw} + S_p + S_c \pm \Delta S_{sw}, \qquad [35b]$$

where V_{iw}, V_{gw}, V_{dw} and V_{tw} are the volumes of irrigation water applied, groundwater influx either by capillarity or direct invasion of the water table, deep percolation of drainage water, and surface runoff (tailwater), respectively; C_{iw}, C_{gw}, C_{dw} and C_{tw} are the soluble salt concentrations in the preceding four types of water, respectively; S_m and S_f are the amounts of salts brought into soil solution by mineral weathering and from the dissolution of fertilizers and amendments, respectively; S_p and S_c are the amounts of salts precipitated out of solution in the soil and removed from the soil solution by crop uptake, respectively and ΔS_{sw} is the change in soil solution salinity (Kaddah and Rhoades, 1976). This relation can be simplified (approximated) for certain situations by making some reasonable assumptions. For example, except for heavy-textured, cracking soils, it may be assumed that C_{iw} and C_{tw} are essentially the same. It may also be assumed, with the shallow water table situation in mind, that C_{dw} and C_{gw} are about the same. Additionally, because S_m, S_f, S_p and S_c are usually small in relation to the other quantities and the contributions of the S_m and S_f inputs tend to offset the losses associated with S_p and S_c, these terms have been traditionally deleted from the calculations (U S Salinity Laboratory Staff, 1954). However, S_p may be significant and some adjustment for this process may be necessary, if salts other than chloride are considered (Rhoades, et al., 1974). Substitution of these "equalities" and assumptions into Equation [35b] yields the following relation, where rainfall is insignificant:

$$\left(V_{iw} - V_{tw}\right)C_{iw} + \left(V_{gw} - V_{dw}\right)C_{dw} = \Delta S_{sw}\,. \qquad [36]$$

This relation, along with measurements of the volume of infiltrated water (V_{iw} - V_{tw}), of the concentration of the irrigation water (C_{iw}), of the concentration of soil salinity at the bottom of the root zone (C_{dw}) and of the change in soil salinity within the root zone over the measurement period, can be used to estimate the net amount of leaching ($V_l = V_{dw} - V_{gw}$) and the load of the salt ($V_l C_{dw}$) draining from the root zone.

Substitution of electrical conductivity for concentration and of the volume of infiltrated water, V_{inf}, for (V_{iw}-V_{tw}) into Equation (36) yields the following equation relating leaching fraction ($L_f = V_{dw}/V_{inf}$) for steady-state ($\Delta S_{sw} = 0$) and good drainage conditions ($V_{gw} = 0$), as was published in Handbook 60 (U S Salinity Laboratory, 1954):

$$L_f = V_{dw} / V_{\inf} = EC_{iw} / EC_{dw}. \qquad [37]$$

For such conditions, the leaching fraction can be obtained from the EC-ratio. Thus, the amount of leaching and salt-load can then be calculated, if EC_{dw} can be measured and if the volume of infiltrated water is known. Similarly, the amount of water consumed by evapotranspiration (V_{cu}) can also be determined, since $V_{inf} = V_{cu}/(1-L_f)$ under such steady-state conditions. Remember that all of the assumptions that are contained in Equation [37] are also inherent in the so-called leaching requirement concept, as traditionally applied. Later, the author will give an example of the use of this relation and show how a correction can be made for the disparity between the concentration ratio and the EC ratio, as well as for the error caused by salt precipitation/dissolution processes.

Rose *et al.* (1979) derived an analogous pair of relations to Equations [36] and [37], respectively, for non-steady-state and for steady-state conditions, meeting the more limiting assumptions described below. The intended use of these relations was for predicting from relatively short term measurements whether leaching would be adequate in the long term to keep salinity within acceptable limits for cropping. The claimed value of this relation was that it would permit *L* (defined below) to be determined from relatively easily obtained information. Retaining most of their symbols, the expression for non steady-state conditions given in differential form is:

$$z\Theta_m\left(\frac{\partial s_m}{\partial t}\right) = Ic - L\, s_z, \qquad [38]$$

where z is soil depth, Θ_m is mean volumetric soil water content averaged over depth z, s_m is the mean concentration of a conservative solute (such as chloride) over depth z, t is time (yr.) measured from an initial time (t_0) when s_m is first measured, *I* is the rate of irrigation averaged over the time period of measurement, c is the concentration of the tracer solute in the irrigation water, *L* is the leaching flux density at depth z averaged over the time period of measurement and s_z is the mean concentration of the tracer solute at depth z over the time period of measurement at the reference water content θ_m.

The following assumptions are inherent in Equation [38]: (i) water and solute flow are only vertical, (ii) the amount of leaching is low and the soils are only slowly permeable, (iii) the tracer-solute is non-adsorbed and non-transformed in the rootzone, (iv) the rates of water application and drainage at an arbitrary specified depth z are constant and equal to their mean annual values, (v) leaching occurs when the soil is essentially saturated, (vi) there is negligible uptake of the solute by the crop and rainfall is negligible, (vii) there is no surface runoff of water, (viii) the irrigations and croppings are uniform within the field, and (ix) the shape of the solute-profiles changes very little over the time period.

Rose *et al.* (1979) also developed the following expressions for predicting the mean solute concentration in the soil profile at later times, including the steady-state condition. For non-steady-state conditions, the Equation is:

$$S_m - S_{m(0)} = \left(I\,c / L\,\lambda - S_{m(0)}\right)\left\{1 - \exp\left[-\left(L\,\lambda / z\,\Theta_m\right)t\right]\right\}, \qquad [39]$$

and for steady-state conditions it is:

$$s_m^* = I\,c / L\,\lambda, \qquad [40]$$

where s_m^* is the mean concentration of the tracer solute in the soil profile at steady-state and $\lambda = s_z / s_m$, which is estimated at any time from the shape of the tracer concentration-profile. The history of s_m can then be computed with the later two equations for all values of t, assuming the irrigation/crop system remains as it was for the period when the data that yielded the value of L was collected. The value of the non-steady-state Equation [39] is that it permits the average annual value of L to be determined from two measured value of s_m and knowledge of I & c over time. Equation [40] is equivalent to Equation [37] when $\lambda = 1$. Equation [38] is analogous to Equation [36].

Slavich and Yang (1990) modified the equations of Rose et al to account for anion exclusion and pore-bypass during leaching. While the refinements contained in their approach to compensate for some of the processes which influence leaching-efficiency are physically sound and potentially beneficial, it will be difficult in actual field practice to obtain the needed inputs for the various parameters, especially for those describing by-pass flow, required by this approach. Furthermore, one would not expect leaching efficiency to be single valued but rather to vary with irrigation systems, with management, with different soil types, with depth in the soil, and from place-to-place in the field. Thus, the method seems needlessly complicated, given the uncertainties in the other inputs and is probably impractical. Reasons for these comments will be more apparent from the information that is given later in this Section. Therefore, their modified relations will not be given herein.

The above relations (Equations [35] to [40]) provide a means for the estimation of the amount of through-drainage (extent of leaching), of the long-term "equilibrium" salinity level resulting from irrigation, and of the salt-loading leaving the root zone (or past the maximum depth of sampling) in different fields of an irrigation project and within the different areas of individual fields. Additionally, they provide a means to locate the primary areal sources of salt-loading from irrigation because, as discussed earlier, the amount of salt-loading is proportional to leaching volume, though it is also affected by V_{inf}, C_{iw} and the properties of the substrata through which the deep percolation flows enroute to its receiving water or soil. Various studies have been undertaken to evaluate the above equations; their findings have been reviewed and critiqued by Rhoades (1997b) and will not be reviewed here. Most of these evaluations were made using analyses (primarily chloride analyses) of **soil samples.** Such analyses are too time-consuming to be very practical for large area assessments of leaching and salt-loading. For this reason some attempts have been made to determine if analogous assessments can be made from the more practical measurements of in-situ, bulk soil electrical conductivity made using the geophysical sensors described earlier, while still utilizing the same salt-balance relations and approach (Rhoades, 1980; Cook *et al.*,1989; Slavich and Yang, 1990).

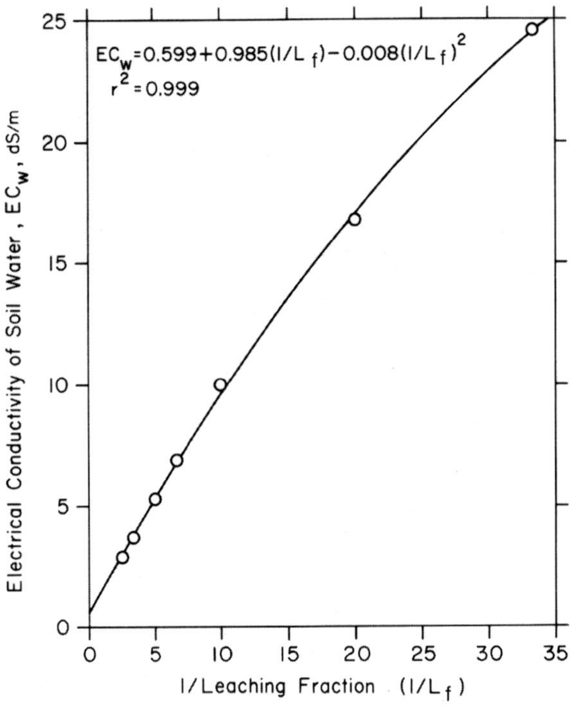

FIGURE 77
Relationship between electrical conductivity of soil water and leaching fraction for the Colorado River water used in the Wellton-Mohawk Irrigation and Drainage District of Arizona, USA (after Rhoades, 1980)

Rhoades (1980) was the first to combine the use of EC_a measurements and salt-balance relations to estimate leaching amounts in irrigated fields. His method is practical and will be described to show how the assessment technology and salt-balance relations described in this book can be combined into a practical procedure for determining the extent of leaching and the areal sources of salt-loading. He used measurements of EC_a (made at the bottom of the root zone with a four-electrode EC-probe), along with soil-specific calibrations relating EC_a and EC_w , the assumption that the EC of the drainwater is the same as the EC of the soil water at or below the bottom of the root zone, and a modified version of the steady-state model (Equation [37]) to infer leaching fraction in some irrigated alfalfa fields in the Wellton-Mohawk irrigation district of Arizona. Equation [37] was modified to incorporate the effects of S_m and S_p (discussed earlier in terms of Equation. [35b]) in the relation between L_f and the EC_{iw} / EC_{dw} ratio, as well as to replace the implied assumption of a 1:1 relation between concentration (C) and EC that is inherent in Equation [37] with a more appropriate curvilinear relation. These modifications were calculated using a steady-state chemical model (Rhoades and Merrill, 1976; Oster and Rhoades, 1990; Rhoades *et al.*, 1992) that incorporates the effects of salt precipitation (S_p) and mineral weathering (S_m) reactions which occur in irrigated root zones on the resulting concentrations of solutes in the soil water and that calculates the corresponding EC value from these concentrations. The following curvilinear relation, which he obtained for the Colorado River water used for irrigation in the Wellton-Mohawk, illustrates the nature of the modification:

$$EC_{dw} = 0.599 + 0.985\left(\frac{1}{L_f}\right) - 0.008\left(\frac{1}{L_f}\right)^2, \qquad [41]$$

with an r^2 value of 0.999 (see Figure 77). The value of EC_{iw} was inherently incorporated into this relation by means of the coefficient values, as are the corrections for salt precipitation and dissolution and for conversion of total salt concentration to its' EC equivalent. These modifications permit EC to be used in place of the conservative chloride solute concentration and Equation [41] used in place of Equation [37] as a basis for estimating the steady-state value of leaching fraction. Additionally, he took steps to assure that the method would apply to each soil type encountered in the surveyed fields. To accomplish this, he established the slope and intercept values in Equation [5] relating EC_a and EC_w for each soil encountered in the project fields (see Figure 20, after Rhoades, 1980). He then substituted Equation [41] into Equation [5] with the assumption that $EC_{wc} = EC_{ws} = EC_{dw}$ (since the measurements applied to the bottom of the rootzone, where $EC_w = EC_{dw}$) to give specific equations relating EC_a and L_f for each of the different soil types encountered in the surveyed fields. These equations took the following form:

$$\left(\frac{1}{L_f}\right) = slope\,EC_a - intercept, \qquad [42]$$

where the slope and intercept values were specific for each soil type (see Figure 78, after Rhoades, 1980). The appropriateness of the approach was tested in four different fields. The leaching fraction at each site was also estimated as the chloride-concentration ratio Cl_{iw}/Cl_{sw}, assuming the chloride ion was a conservative (tracer) solute that would behave ideally according to Equation [37]. The steady-state assumption was assumed applicable for the test-fields because the same crop had been irrigated for many years in each field with consistent management. This assumption was concluded to be appropriate, since essentially the same values of L_f were determined using both methods (Equation [42] and chloride ratio) of estimation (see Figure 79, after Rhoades, 1980). It was also concluded from these results that the more practical measurements of EC_a could be used in place of the soil sampling/chloride ratio procedure to estimate leaching fractions. Since the composition of the drain water is defined in relation to either the leaching fraction value or EC_a value in this approach, it is a simple matter to express the data in

FIGURE 78
Calibrations established between leaching fraction (L_f) and electrical conductivity of soil (EC_a) for different soil types in the Wellton-Mohawk Irrigation and Drainage District of Arizona, USA (after Rhoades, 1980)

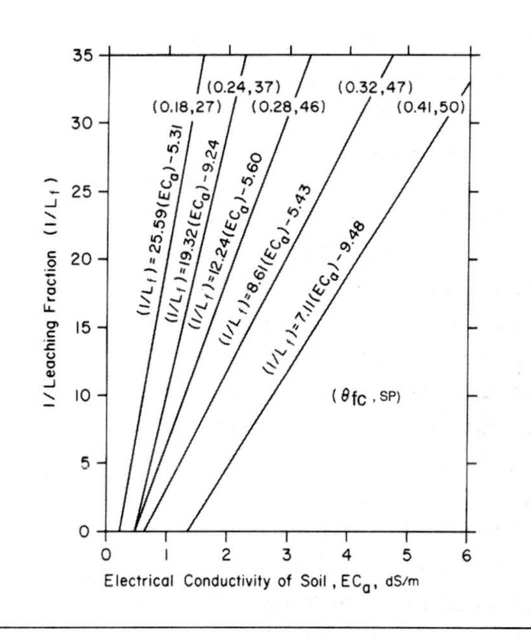

FIGURE 79
Correlation found between electrical conductivity of soil water (EC_w) and the ratio of chloride concentration in soil water (Cl_w) below the rootzone to that in the Colorado River irrigation water for the Indio fine silty loam soil at four study sites in the Wellton-Mohawk Irrigation and Drainage District of Arizona, USA (after Rhoades, 1980)

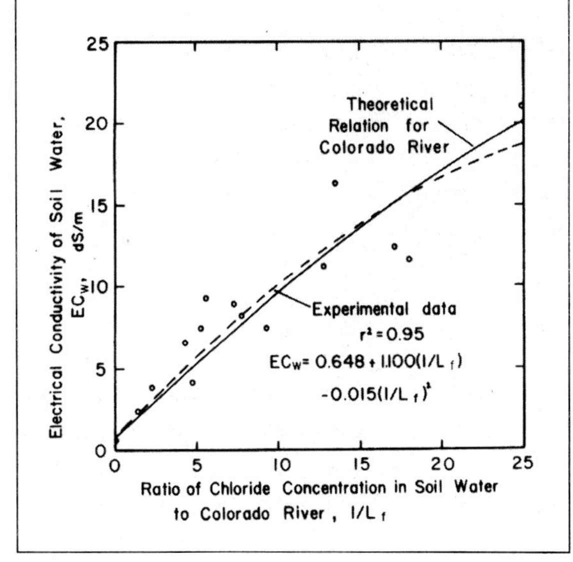

terms of absolute salt-load, provided V_{inf} is known. This latter information, however, will be difficult to obtain in practical field appraisals, especially at the many different sites in the field where measurements of EC_a are easily made. Analogously, the areal sources of salt-loading and the regions of high leaching can be readily inferred and delineated by representing the data in the form of a leaching fraction map. Of course, inherent to the success of this approach are the following assumed conditions: (i) the soil is at steady-state, (ii) the chemical composition of the irrigation water is known and essentially constant over time, (iii) the soil water is near field-capacity at or below the lower extremity of its root zone, (iv) the relation between soil electrical conductivity and soil water electrical conductivity is known for the soil type that exists at that depth and field-site, and (v) there is no upward flux of water and salt from the shallow water table into the depth of EC_a measurement. For cases where the latter condition exists, or other such non steady-state effects exist, an analogous approach based on Equation [36] or [38] could be developed to estimate leaching and salt-loading amounts. The advantage of this spatially-calibrated EC_a-sensor approach is that a detailed pattern of leaching and salt-loading, and of their variability within the field, can be more practically obtained using the mobilized system of calibrated-EC_a measurements described earlier than is possible using the soil sampling/laboratory procedures employed by the preceding investigators. Additionally, the salt-load can be more accurately estimated using the approach of Rhoades (1980), because the chemical composition of the drainwater is known as a function of L_f by means of the chemical model which is incorporated into the approach.

The above example, as well as those of Cook *et al.* (1989) and Slavich and Yang (1990), studies have demonstrated that a generally good potential exists for the assessment of leaching rates in fields and for their associated salt-loads using relatively simple salt-balance models and geophysical sensors which measure soil electrical conductivity. This is especially true, if these models are modified by a method analogous to that described above so as to correct for some of the errors associated with the simplifying assumptions inherent within them. However, no one to date has combined all of these refinements together with the measurement/calibration methods into a practical system for such an assessment, especially one that permits the infiltration amounts to be established at the many points of measurement that are needed to account for the highly variable conditions of irrigation and leaching that typically exist in agricultural fields. Some data illustrating the degree of this variability and of the need to account for it in the desired assessments will be presented and discussed in the following paragraphs; additionally, some procedures will be suggested to make the assessments more practical than was possible when many of the previously discussed methods were developed.

The practicality of the application of the salt-balance approach to the estimation of leaching and salt-loading rates has been markedly enhanced through the development by Rhoades and collaborators of the integrated, mobilized system of soil electrical conductivity measurement and of the salinity calibration software that was described earlier. This technology now makes it practical to obtain detailed accurate information of soil salinity distributions in irrigated root zones and fields. This technology has already been demonstrated to be capable of providing useful, qualitative information about irrigation uniformity, the adequacy of drainage, the relative degree of leaching, and the major source areas of excessive deep percolation and salt-loading (Rhoades *et al.*, 1997a; 1997b). The utilization of this technology in combination with improved salt-balance models offers good potential to make rapid quantitative estimates of leaching and salt-loading rates. Some examples will now be given to support these conclusions and to illustrate the utility of this technology.

Many of the examples already given in the section *Evaluating adequacy and appropriateness of irrigation/drainage* demonstrate how the relative degree of leaching can be inferred from the level and pattern of soil electrical conductivity within the root zones of irrigated fields. Figure 67 illustrates the often observed relatively high leaching that occurs in the "upper" sections of furrow-irrigated fields, as does Figure 68. With reference to Figures 70 and 71, it was shown how the net leaching fraction could be inferred from the mean salinity level of the root zone and chemical-model (WATSUIT) predictions assuming steady-state conditions. Additionally, these data and those of Figure 72, show how variable leaching can be from place to place within individual fields and how useful the mobilized systems of salinity assessment can be to determine this variability, its patterns and causes. With reference to Figures 74 and 75 and Tables 4 and 5, it was shown how the shape of the salinity profile could also be used to advantage for ascertaining and mapping those regions of the field where the relative leaching flux was excessive, or inadequate, as well as for evaluating the suitability of the irrigation practices. The absolute amount of leaching can be inferred from such data provided the amount of infiltration can be established for the various sites, and in turn the associated salt-load can be calculated from the EC_{dw} , which can be determined from the assessment measurements by the methods previously explained/demonstrated. Practical methods are not available to measure this spatial variation in infiltration amount. Hence, the levels of soil salinity can be used to estimate the distribution of leaching rates that have occurred within the field from the overall field-average of infiltration amount and , in turn, the distribution of the associated salt-loading from the rootzone. Such methodology needs to be implemented to achieve management that is efficient in water use and protective of the environment. All of the measurement aspects needed to do this have been presented in this report.

SCHEDULING AND CONTROLLING IRRIGATIONS

Practical means of scheduling and controlling irrigations to conserve water while avoiding yield loss from salinity or water deficiency have not been given much attention. Presently utilized typical methods of scheduling irrigations are based on measurements of soil water status and/or predicted evapotranspiration. These latter methods are inadequate for saline soil conditions because they do not take into account salt (osmotic) effects on water availability, which also depends on soil water depletion allowed between irrigations. Furthermore, direct measurements of soil water depletion or matric potential can not be used to control the leaching fraction which is required to prevent excessive soil salinity accumulation. For saline water, irrigations must be scheduled before the total soil water potential (matric plus osmotic) drops below the level which permits the crop to extract water at a sufficient rate to sustain its physiological processes without excessive loss in yield. The crop's root system normally extracts progressively less water with increasing soil depth because rooting density decreases with depth and because available soil water decreases with depth as the salt concentration increases (Rhoades and Merrill, 1976). Therefore the frequency of irrigation should be determined by the total soil water potential in the upper rootzone where the rate of water depletion is greatest. On the other hand, the amount of water to apply depends on stage of plant development and the salt tolerance of the crop and, consequently, should be based on the status of the soil water at deeper depths. In early stages of plant development it is often desirable to irrigate to bring the soil to "field capacity" to the depth of present rooting or just beyond. Eventually, however, water must be applied to leach out some of the salts accumulating in the profile to prevent salt concentration from exceeding tolerable levels. Thus, the amount of water required is dictated by volume of soil reservoir in need of replenish-ment and level of soil salinity in the lower root zone.

Since soil electrical conductivity is a tracer of the interactions of water infiltration, evapotranspiration and leaching as demonstrated above, it can be used as the basis for irrigation/salinity management. As shown earlier soil water salinity (and hence osmotic potential; M Pa at 25 C° ≅ 0.04 x EC_{25}, in dS/m) and leaching fraction can be determined from measurements of soil electrical conductivity. Also as shown earlier, EC_a is, for a given soil, mostly responsive to the EC (hence osmotic potential) of the soil water in the pores which supply most of the water to the plant and it can be used to determine leaching fraction. With calibration for the particular soil-type, the total soil water potential can be determined from EC_a and Θ_w using Equation [5] and knowledge of the matric potential-Θ_w relation for the soil and the leaching fraction can be determined solely from EC_a. The means for this and data showing its feasibility have been demonstrated by Rhoades *et al.*(1981).

Rhoades *et al.* (1981) showed/concluded that the EC_a measurements could be made in the upper profile to schedule irrigations and the latter measurements could be made in the lower profile to determine that sufficient, but not excess, water is being applied over the long-term to keep salinity within acceptable limits. As also shown by Rhoades *et al.* (1981), one can also associate a "set-point" value of EC_a (equivalent to a desired total water potential) to use as a basis for scheduling irrigations, but the combined use of moisture and salinity measuring sensors would likely be more

FIGURE 80
Water penetration following a small irrigation, as deduced from measurements made with four-electrode sensors and a neutron probe (after Rhoades *et al.*, 1981)

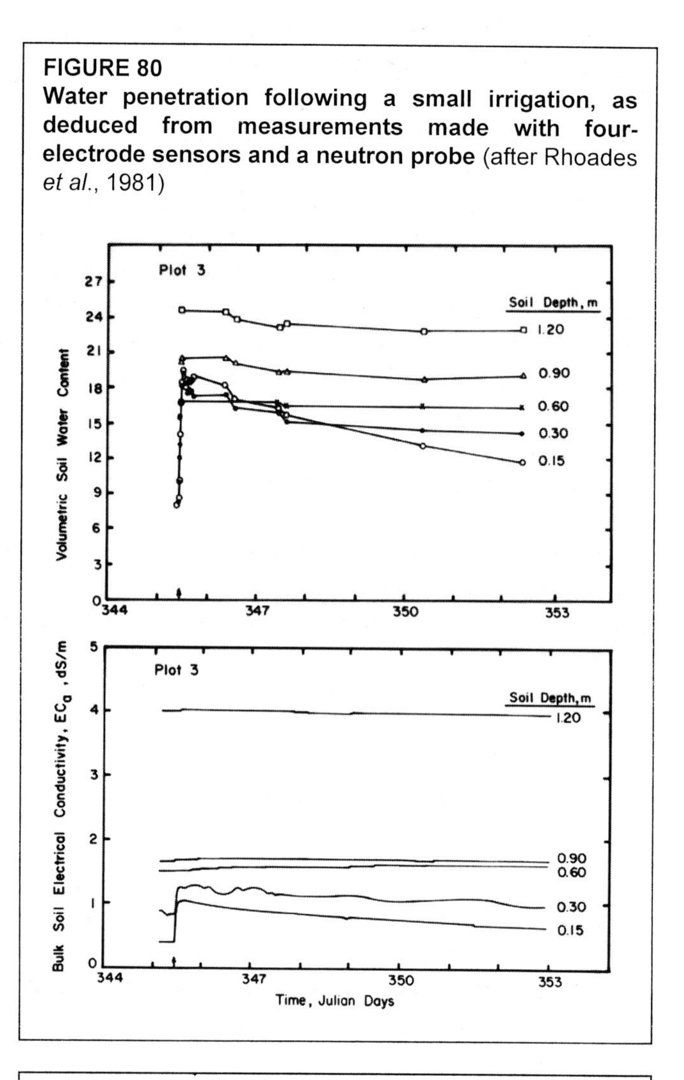

FIGURE 81
Water penetration following an irrigation of moderate amount, as deduced from measurements of soil electrical conductivity (EC_a) with four-electrode sensors (after Rhoades *et al.*, 1981)

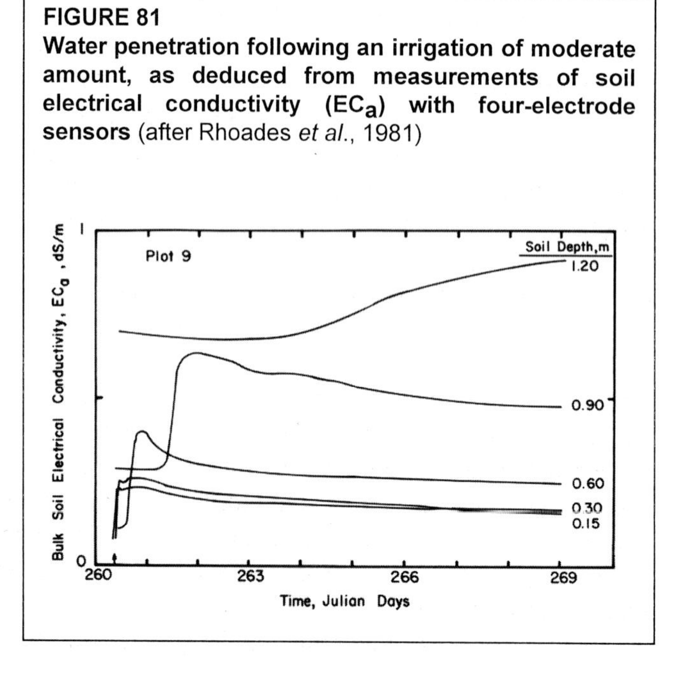

accurate (evidence of this is shown in Figures 22 and 80). For such scheduling and monitoring purposes, the measurements of EC_a can be made at monitoring locations using burial-type four-electrode sensors or estimated at many places in the field using the above-ground sensors and profile estimating procedures, as described earlier.

At the end of an irrigation cycle, a certain EC_a-depth relation will exist through the soil profile for any given combination of soil-, plant- and water-type. Upon irrigation, the water content will increase at every depth in the profile reached by the wetting front and EC_a will increase correspondingly as water flows into that soil volume and Θ_w increases (see Figures 80 and 81). Thus, in principle, irrigations can be automated using EC_a-sensors placed in the profile at desired depths to initiate irrigations when the set point value is reached and to terminate them when the EC_a reading at the desired depth shows the arrival of water and/or sufficient leaching to keep salinity within limits. An example of the use of EC_a-depth measurements to sense the movement of an irrigation wetting front is shown in Figure 81.

The theory and data presented here and in Rhoades, et al. (1981) support the conclusion that measurements of soil electrical conductivity could be used to schedule irrigations, control the depth of water penetration, and obtain the desired leaching fraction. Some irrigation systems could be automated with burial-type sensors. The methodology has good potential that should be exploited where salinity is a limiting factor.

Reclamation of Saline Soils

Though excessive levels of salts in soils can sometimes be reduced over time using management practices that are compatible with cropping, it is more usual to set aside cropping temporarily and to speed the removal process by reclamation practices. The selection of appropriate reclamation practices requires knowledge of the cause and source of the salt-related problems. Some examples were given in earlier sections to illustrate the utility of the salinity assessment methodology to help determine such causes and sources.

The only practical way to reduce excessive soluble salts in soils is to leach the salts out by passage of lower-salinity water through the active rootzone depth of the soil. The amount of leaching required to reclaim saline soils is a function of the initial level of soil salinity, of the level desired and the depth of soil needed to be reclaimed (which are largely determined by the crops to be grown), and of certain soil and field properties and the method of water application (which influence leaching efficiency). Several theories have been developed to predict needed leaching but various required parameter "unknowns" usually limit their usefulness and accuracy without on-site calibration. For this reason empirical relations, which are based on field experiments or experience, are generally used as guidelines for reclamation. The accuracy of these guidelines are unknown for most situations.

A simple, straight-forward way to determine the amount of leaching required for saline soil reclamation for a particular field and method of water application, or to follow its rate of accomplishment, is to initiate leaching of a test site by the intended method of water application and to follow the change in soil salinity with amount of water application/infiltration. A very convenient, practical monitoring approach is to measure changes in soil electrical conductivity, in this regard, using any of the sensor techniques previously described (though a burial-type four-electrode EC-probe would be more generally appropriate). The progress of salt removal is immediately evidenced from the EC_a readings during leaching; this clearly and simply establishes the required leaching for that field and water application condition. The data shown in Figure 82 (after Rhoades, 1979b) illustrate and demonstrate this method for a simple situation where a "salinity-probe" was installed at a depth of 15 cm for monitoring purposes. Analogous data would be used to monitor the deeper depths and rate of progress.

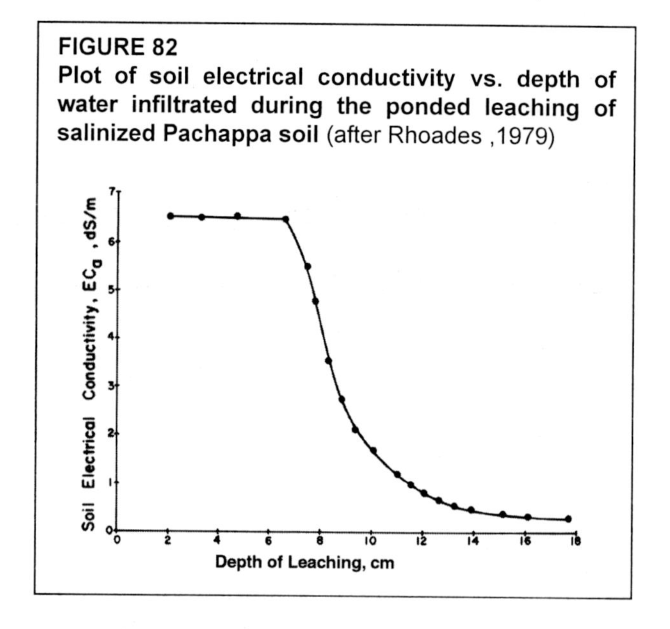

FIGURE 82
Plot of soil electrical conductivity vs. depth of water infiltrated during the ponded leaching of salinized Pachappa soil (after Rhoades ,1979)

The process of detecting the degree of change resulting during reclamation process is the same as that previously described for monitoring salinity; hence, it will not be repeated. The changes resulting from the reclamation processes in the mean value of salinity and in its spatial pattern for a given field would be determined by the same set of calculations provided in Annex 9.

Chapter 5

Operational and equipment costs associated with salinity instrumentation measurement techniques

Salinization is the increase in concentration of total dissolved solids in soil and water. Secondary salinization is the terminology commonly used for any type of human activity which increases the salinization of land or water resources. Secondary salinization of land resources has been occurring since the beginnings of human settlement, and is most often directly related to the development and expansion of irrigated agricultural practices.

Secondary salinization reduces agricultural crop yields, degrades land values, and if left unabated, eventually leaves the affected soil in an unusable state. The economic costs of secondary salinization at both the farm and regional scale have been well documented. Grieve *et al.*, explored the economic costs of waterlogging and soil salinity to the Murray Valley basin in New South Wales, Australia (Grieve *et al.*, 1986). They estimated these costs to be in excess of 16 million Australian dollars on an annual basis, which represented 16% of the district's total agricultural production. In a follow up study in 1994, Oliver *et al.*, performed a capital cost survey of 177 local government agencies and 39 public utilities located in the greater Murray-Darling Basin (Oliver *et al.*, 1996). They found that over 27.7 million Australian dollars were allocated and/or distributed for salinity related repairs and maintenance, and for salinity related research and education. In 1994, Luke and Shaw performed economic assessments of the costs of salinity to agriculture in the Loddon, Campase, and Avoca dryland sub-regions of Australia (Luke and Shaw, 1994a,b,c). They found that if left unabated, by the year 2001 the total economic losses for salinity in these three regions could reach 880, 323, and 531 thousand Australian dollars, respectively.

The total area of salt-affected land on a global basis has been estimated to be approximately 76.3 million hectares (Mha), of which 41.5 Mha is considered to be seriously degraded (Oldeman *et al.*, 1991). Serious secondary salinization is occurring at an ever increasing rate across the world's irrigated agriculture, and is responsible for substantial economic losses in agricultural production. Overall, the ratio of salt-affected to irrigated land has been estimated to be between 9 to 34% in the following countries: Argentina, 33.7%; Australia, 8.7%; China, 15%; Commonwealth of Independent States, 18.1%; Egypt, 33%; India, 16.6%; Iran, 30%; Pakistan, 26.2%; South Africa, 8.9%; Thailand, 10%; and in the United States, 23% (Ghassemi, Jakeman, and Nix; 1995).

Secondary salinization can be effectively controlled, provided proper land management and agricultural production methodologies are employed (Rhoades, 1997). However, the successful implementation of agricultural salinity management strategies depends upon many political, economic, and technical factors which must be synthesised together in an objective fashion. In general, this process is not possible unless the magnitude and distribution of the soil salinization can first be quantified. Therefore, the inventorying and/or monitoring of soil salinity in a cost effective manner represents a critical first component to this management process.

Within the last 25 years, a tremendous amount of progress has been made in regard to the assessment of soil salinity using electrical conductivity measurement techniques. These survey instrumentation techniques have been shown to be both highly accurate and rapidly employable, and in general represent the most cost effective salinization inventorying methodologies currently in use. However, to date these techniques have not been fully realized nor taken advantage of.

In order to objectively quantify the cost benefits of the above instrumentation methodologies, an economic cost analysis must be performed. Such an analysis must include two components; (1) a detailed description of the capital costs associated with the various equipment used in the most common survey instrumentation techniques, and (2) an assessment of the operational costs incurred when applying these techniques for the measurement of soil salinity.

SALINITY INSTRUMENTATION: EQUIPMENT SPECIFICATIONS AND COST INFORMATION

Before an operational cost analysis can be performed on any type of soil salinity survey instrumentation technique, the performance specifications and capital costs of the instrument must first be determined. Hence, a summary of equipment specifications and product cost information for the most commonly used field salinity assessment instruments is given below. This discussion includes information concerning both manual (hand-held) instruments and mobilized salinity survey systems, global positioning systems (GPS), relevant analytical equipment, and soil salinity assessment and mapping software.

Soil Salinity Survey Instruments

Non-invasive Electromagnetic Induction Instruments

EM38

The Geonics EM38 was designed specifically for agricultural soil salinity surveys, and can be used to survey large areas quickly without employing ground (contact) electrodes. The EM38 uses Geonics patented electromagnetic induction principle, providing depths of exploration of 1.5 meters and 0.75 meters in the vertical and horizontal dipole modes, respectively.

The EM38 is very lightweight (2.5 kg), compact (1 meter long), and highly durable. It can be used to measure either apparent conductivity in millisiemens per meter (mS/m) or the inphase ratio of the secondary to primary magnetic field in parts per thousand (ppt). Either conductivity or inphase measurements can be collected in both the horizontal and vertical dipole modes. Measurements are normally made by placing this instrument on the ground and physically recording the meter reading. Signal readings can also be digitally logged by using a data logger in conjunction with the meter (see DL720 Data Logger, described below).

Measurements can be made either manually (using a trigger switch) or in a continuous mode. Both the meter and data logging system can be operated by a single surveyor.

The EM38 is well suited for mechanized applications, since it can be easily mounted to various types of transport vehicles and/or towing platforms and its data acquisition software can be readily modified to interactively communicate with various types of computer systems, controller boards, and/or other survey instrumentation (such as GPS survey systems, etc.).

DL720 Digital Data Logger

The DL720 digital data acquisition system is a rugged and weatherproof logging system designed to support all Geonics ground conductivity meters. The DL720 system includes an Omnidata Polycorder 700 series data logger, interconnecting cables, and DAT software for data storage and manipulation.

Conductivity data can be collected and stored in either static (manual) or continuous mode (using a time interval selected by the surveyor). Stored data are downloaded from the data logger directly into the DAT software program using RS232 serial connections. All Geonics DAT software has been designed for use on IBM compatible personal computers, and can be used to edit, plot, and print profiles of either the conductivity or inphase signal data. The data logging software also includes a time-stamp option, which facilitates the merging of any DAT data file with many common GPS file formats.

Invasive (in situ) Electromagnetic Induction Instruments

SCT-10 Conductivity Monitoring System

The Martek Model SCT-10 is a portable, monitoring system capable of *in situ* measurement of soil conductivity and temperature. The SCT-10 is small, light weight, and contains its own rechargeable battery power supply, internal clock, memory, and communications port. All meter calibration and computations are performed via a low power, CMOS microprocessor while a large, alphanumeric display provides instructions and data in simple English.

The SCT-10 accepts a wide assortment of conductivity sensors for soil and water applications and is capable of accepting all common cell constants and adjusting any collected conductivity data to 25 degrees centigrade. Signal data measurements are displayed in direct engineering units, and simultaneous digital and analog signals are available for secondary recording instruments. For recording data, the SCT-10 comes standard with 0-1 volt DC for analog recorders and a serial ASCII port for communication with computers or controlling devices.

Four types of soil sensors are available for use with the SCT-10; including a (i) vertical sensor, (ii) horizontal array, (iii) bedding sensor, and (iv) burial sensor. The vertical sensor (commonly referred to as a standard soil-probe, insertion four-electrode probe, or Rhoades probe) can be used to acquire conductivity readings down the first 1.2 meters of the soil profile. The bedding probe is a miniature version of the vertical sensor, and is designed primarily for acquiring near-surface conductivity readings in seed beds or high density root areas. The burial sensor is an in situ sensor which can be left buried in the ground to facilitate the long term monitoring of soil conductivity over time. The last soil sensor, the horizontal surface array, can be used to determine the average conductivity of large volumes of soils by manually varying the configuration of the current electrodes in the array.

Equipment Specifications and Product Cost Information

Detailed equipment specifications and product cost information for the Geonics EM38 meter and DL720 digital data logger are listed in Table 6. Equipment specifications and cost information for the Martek SCT-10 meter and sensors are given in Table 7.

TABLE 6
Geonics EM38 and DL720 equipment specifications and costs.

EM38	
Measurements	conductivity (mS/m) or inphase ratio (ppt)
Sensor	dipole transmitter
Intercoil spacing	1 meter
Operating frequency	14.6 kHz
Power supply	9 volt alkaline battery
Conductivity range	100 to 1000 mS/m
Inphase range	±29 ppt
Resolution	±0.1% of full scale
Accuracy	±5% at 30 mS/m
Instrument dimensions	103 x 12 x 12.5 cm
Case dimensions	117 x 19 x 13 cm
Instrument weight	2.5 kg
Shipping weight	10 kg (including case)
Cost	US$7 195
DL720 Specifications	
Storage capacity	16 500 1-channel records; 10 000 2-channel records
A/D resolution	16 bits
Dimensions	20 x 10 x 5.3 cm
Weight	1.5 kg
Cost	US$ 3 775 (includes DAT software and interconnecting cables)

TABLE 7
Martek SCT-10 equipment specifications and costs.

Measurements	temperature (°C) and conductivity (milli S/cm)
Sensor types	vertical, horizontal, bedding, in-plant (in situ), flow-through (liquid)
Power supply	rechargeable, internal nicad battery
Conductivity range	0-1, 0-10, 0-100 milli S/cm
Temperature range	0-50 °C
Resolution	±0.0001, ±0.001, ±0.01 milli S/cm; ±0.01 °C for temperature
Accuracy	±0.01, ±0.05, ±0.5 milli S/cm; ±0.1 °C for temperature
Instrument dimensions	16.5 x 10.1 x 20.3 cm
Instrument weight	2.7 kg (not including sensors)
Cost (add 30% for foreign orders)	SCT-10 meter: US$ 2,000 Probes: (i) Vertical probe: US$500; (ii) Horizontal probe: US$300; (iii) Bedding probe: US$250 and (iv) Burial probe: US$100.

GPS Equipment

Other useful instruments for soil salinity survey work include, but are not limited to, soil temperature probes (for measuring soil temperature throughout the soil profile), GPS survey equipment (for acquiring spatial location), laser range finding systems (for measuring changes in micro-elevation), and various sensors designed to measure other types of soil or crop attributes (such as infrared spectrometers for measuring crop biomass or time domain reflectometric sensors for measuring volumetric soil water content). Amongst all these instruments, a reliable and versatile GPS system is by far the most important.

There are three primary reasons why a good GPS system should be employed when performing soil salinity survey work. First, salinity survey data are almost always spatial in nature; i.e., survey data is generally acquired across a spatial region and hence the spatial locations of the survey sites must be known in order to construct any sort of contour or relief map. While it is possible to physically grid (i.e., measure) out these survey locations, this data can typically be acquired faster and more efficiently using a GPS system. Second, nearly all types of mechanized conductivity transport systems collect their signal data "on-the-go"; i.e., the conductivity data is acquired while the vehicle is continuously moving. This in turn requires the use of a GPS system, since this is the only type of commercially available system which can readily collect and record the continuously changing spatial location of a moving object (i.e., the moving transport vehicle). And third, nearly all commercial GPS units are designed to interface with multiple types of geographic information systems (GIS). Thus, the transfer of the spatial location information into a GIS is readily facilitated using GPS equipment.

Typical GPS systems range in price from a few hundred to many thousands of dollars or more, depending on the system accuracy specifications. Most types of salinity survey work require survey location accuracies of ±1 to 2 metres, which can be obtained from GPS units which facilitate differential correction (the typical cost of a differentially correctable GPS unit starts around US$1000). Differential correction can be obtained in two ways; either through post-processing or real-time. Post processing (i.e., correcting the data after it has been collected and downloading from the GPS unit) is usually more accurate and less expensive. However, real-time differential correction is generally required when one must navigate to pre-determined survey locations.

The largest commercial GPS companies and satellite differential correction suppliers include Trimble Navigation, Magellan, Garmin, Ashtech, Racal, and Ominstar. These companies can be contacted directly for current product line and pricing information.

Mobilized Soil Salinity Survey Systems

Mobilized Systems for Non-invasive Instruments (EM38)

In its simplest form, the mobilization of the EM38 can be achieved by simply mounting the instrument on some type of sled or trailer and then towing it over the survey area. The sled or trailer must be completely non-metallic and must be kept at least 1 to 2 meters from the towing vehicle (which can be either a small tractor or an alternative terrain vehicle). A GPS system must also be set up and mounted on the towing vehicle (or more preferably, to the trailer itself) in order to simultaneously record spatial location information. The EM38 would be set to record conductivity data in a continuous mode (defined by a time interval which may be selected by the surveyor) and this data would then be automatically "time-stamped". Hence, once the survey was completed, this data could be readily merged with most types of post-processed GPS location information to create a spatially referenced conductivity map.

There are currently no third-party vendors who sell commercially available, "off-the-shelf" sleds or trailers for the EM38. However, these types of towing apparatuses are fairly simple to fabricate, and can generally be built for between US$500 to US$3 000 by many trailer fabrication shops.

More complex trailering platforms generally require data logger software modification and/or customized system integration. For example, a simple EM38 trailer design will typically support (i.e., physically hold) only one EM38 unit at a fixed height and orientation above the soil surface. To acquire both horizontal and vertical EM38 signal data, the trailer must support and control two EM38 units (operated in sequence), or the trailer must be able to mechanically rotate the unit (as well as control the timing of the data logging). Additionally, it is usually desirable to directly interface the GPS system with the DL720 data logger so that either system can communicate (and/or control) the other. Again, there are currently no commercially available systems which are designed to perform these sort of survey operations. However, there are geophysical companies which can perform the software modification and/or system integration necessary to create such trailering platforms. One such company with previous commercial experience in Geonics DL720 custom software modification and EM38/GPS system integration is Geomar Geophysics Ltd., in Toronto, Canada. Geonics Limited can also sometimes recommend one or more appropriate companies for the design, integration, and fabrication of advanced trailering platforms.

The total cost of such a trailer obviously depends on the mechanical and electronic complexity of the platform design. Hence, estimating an approximate cost for such a trailering system is impossible without first knowing the design specifications. However, a very gross price range for most types of complex platforms would be between US$2 500 to US$10 000. (Note that this cost does not include the EM38, GPS system, or towing vehicle.)

An even more sophisticated and versatile approach to mobilized surveying can be employed by suspending and controlling the EM instrument(s) directly from a specialized, self contained transport vehicle. This sort of system has been developed at the United States Salinity Laboratory by Rhoades and colleagues (Rhoades 1992a, 1992b, 1993, 1994, 1996b; Carter, et al., 1993) and is currently being developed for commercial distribution by Agricultural Industrial Manufacturing (AIM) Inc., in Lodi, California. This automated transport system incorporates both the use of a Geonics EM38 meter and modified Martek SCT-10 meter / horizontal array system, along with a synchronized GPS system for recording spatial location information. This automated transport system can also be readily adapted to incorporate additional sensors. Retail pricing information for this system is not yet available; however, Table 8 lists the price breakdown of the various costs associated with fabricating the basic hydraulic transport vehicle (available from West Texas Lee Company Inc.) into a multi-purpose salinity assessment vehicle.

TABLE 8
Price breakdown for fabricating a basic hydraulic transport vehicle into a multi-purpose salinity assessment vehicle.

Design and fabrication category	Modification costs, US$
1. Construction of front mast assembly with EM38 tube and rotating unit	5 883
2. Horizontal array probe assemblies and scissors frame structure	5 900
3. Electronic controls, computerized control system and programming design	6 450
4. Wiring, limit switches, enclosures, and control box	2 680
5. Hydraulic control valves, manifold, and plumbing	2 560
6. Console, mounting brackets, and accessories	1 490
7. Parking brakes on drive wheels, positive neutral on hydrostatic pump	1 980
8. Miscellaneous: bracket, paint, etc.	990
Total modification cost	**27 933**

Note: total modification cost does not include the costs of the EM38, DL720, SCT-10, GPS, or cost of the basic hydraulic transport vehicle.

Mobilized Systems for Invasive Instruments

Like the non-invasive EM instruments, mobilized versions of conductivity sensors using four-electrode technology have been developed for both research and commercial applications. The most common systems employ one or more sets of four electrodes (in the form of either penetrating shanks or circular disk blades) which are mounted to either a tractor tool bar or trailer platform. This tool bar or trailer platform is then towed or dragged across the survey area, allowing for the near continuous collection of conductivity data (up to one reading per second). These systems are also typically designed to interface with and/or incorporate simultaneously recorded GPS location information.

A mobilized, tractor-mounted version of the horizontal fixed-array conductivity sensor system has been developed at the United States Salinity Laboratory by Rhoades and colleagues (Rhoades 1992a, 1992b, 1993, 1994, 1996b; Carter *et al.*, 1993) and is also currently being developed for commercial distribution by AIM Inc., in Lodi, California. Retail pricing information for this system is not yet available; however, Table 9 lists a price breakdown for assembling such a sensor system. A commercial version of a similar four-electrode system (employing the use of circular disk blade technology) is currently available from Veris Technologies in Salina, Kansas. This latter system simultaneously collects two sets conductivity data by employing two sets of horizontal four-electrode arrays, and can be towed using an alternative terrain vehicle (ATV) or standard tractor. However, this system is not designed to be operated in a bed-furrow environment (the spacings between the disk blades are not adjustable and the clearance ratio between the toolbar and the soil surface is less than 30 cm). More detailed equipment specifications and product cost information for the Veris 3100 System are given in Table 10.

TABLE 9
Estimated system equipment costs for building a mobilized, horizontal array conductivity sensor system

Component	Estimated costs, US$
1. SCT-10 unit	2 000
2. SCT-10 modifications (includes pre-amp, RS232 serial interfacing, and circuit board modifications)	1 000
3. Tractor mounted tool-bar with adjustable vertical penetrating shanks	2 000
4. Wiring and instrument control box	250
Total system cost	**5 250**
Note: total estimated system cost does not include the cost of the GPS system or tractor.	

TABLE 10
Verris 3100 system equipment specifications and costs

Item	Cost, US$
Verris 3100 Soil mapping system. with standard features: - all-welded tubular steel frame - heavy duty spring-loaded coulter/electrodes - non-invasive 17 inch flat disk blades - P205 R75 highway tires - ratchet raising/lowering system - adjustable 4-position clevis hitch - built in micro-processor - 3.5 inch floppy disk drive - internal flash memory to store up to 10 hours of data - back-lit transflective display - DGPS compatible RS232 serial port	11 000
Road kit: ball hitch, lights	175
Weight package	725
Notes: (1) Conductivity measurements acquired in milli S/m; supplier should be contacted directly for conductivity range, resolution, and accuracy specifications under various operating environments. (2) Verris 3100 system cost does not include GPS or towing vehicle costs	

Analytical (Laboratory) Conductivity Instruments

There are numerous analytical conductivity meters which are capable of measuring the conductivity of a solution extract. The cost of such equipment can be quite variable, depending upon the specific instrument specifications. However, there is one conductivity meter which deserves special mention because of its unique ability to measure soil conductivity in a saturated soil paste; the Hach CO150 Conductivity Meter.

The basic Hach CO150 system includes a portable battery operated meter, one conductivity probe for solution extracts, and one soil cup for measuring conductivity of a saturated soil paste. The CO150 system also includes all the necessary software to estimate soil salinity from the saturated paste conductivity reading, based on the methodology of Rhoades et al., 1989b,c. This latter ability makes the CO150 system uniquely different from all other commercially available conductivity meters, since it is the only meter to incorporate this technology. The basic CO150 system is scheduled for commercial release in early 1998. (An upgraded system is also scheduled for release in 1998 which will include a sodium probe, pH probe, and additional software for estimating the sodium adsorption ratio from the saturated paste conductivity, sodium, and pH readings.) Advanced equipment specifications and retail price information for the basic conductivity system are given in Table 11.

TABLE 11
Equipment and estimated cost specifications for basic Hach CO150 conductivity meter

Measurements	temperature (°C), conductivity (milli S/cm) and total dissolved solids (mg/L)
Electrode types	conductivity probe (liquid), soils cup (saturated paste)
Power supply	9 volt alkaline battery (9 VDC line adapter to 115 or 230 volts also available)
Conductivity range	0-0.2, 0.2-2, 2-20, 20-200 milli S/cm
Temperature range	-10-110 °C
TDS range	0 to 19 900 mg/L
Resolution	3 significant digits in conductivity or TDS mode ±0.1 °C for temperature
Accuracy	±0.5% of full scale within each range of conductivity ±1.0 °C for temperature ±1% RSD, 5 to 70 °C for TDS
Instrument dimensions	20.5 x 8.3 x 4.8 cm (not including probe or soils cup)
Data logging	50 data set storage; LCD display output and RS232 output
Cost	US$1 295 to US$1 895(estimated, includes one probe, soils cup, and software)

There can be significant cost savings associated with the measurement of soil salinity in the saturated soil paste, as opposed to a solution extract. This is due primarily to the speed in which samples can be processed and the ability to forego the use of a vacuum extraction system. Additionally, saturated soil paste conductivity measurement systems make it simpler and more inexpensive to process salinity samples "in house", and thus reduce (or eliminate) the need to send the soil samples out for commercial laboratory analysis. The various costs associated with commercial versus internal laboratory sample analyses are discussed in detail in the section *Operational costs associated with the appraisal of soil salinity.*

Soil Salinity Assessment and Mapping Software

All instrumental salinity assessment methods used for field survey work require that measured conductivity readings be somehow converted into (estimated) salinity levels. This conversion can be made using either a deterministic (theoretical) or stochastic (statistical) model. In either case, computer programs can be employed to perform the necessary conductivity- to-salinity calculations, and/or generate the estimated salinity map.

All of the formulas needed to estimate soil salinity from bulk soil conductivity data in conjunction with known or estimated secondary soil properties is given in Rhoades et al., (1989a). Statistical software for estimating spatial regression conductivity-to-salinity models (given a limited set of calibration soil samples) is available from the United States Salinity Laboratory (the ESAP Software Package, version 1.0; Lesch *et al.*, 1995c). ESAP is public domain software, and available free of charge.

Numerous high quality contouring programs are available commercially. Any professional grade mapping and contouring package can be employed for the purpose of creating two or three dimensional conductivity and/or salinity maps. Retail prices for such software typically range between US$250 to US$750.

Company Information

Full addresses are given in Table 12 for all the salinity instrumentation companies discussed in this report.

OPERATIONAL COSTS ASSOCIATED WITH THE APPRAISAL OF SOIL SALINITY

To appreciate the potential cost savings which can be realized from employing survey instrumentation techniques, the overall operational costs for these various methods must be first determined and then compared to the costs associated with conventional sampling. Such an analysis will invariably depend on a number of economic and/or technical assumptions, and these assumptions should be clearly stated and explained. However, there are two underlying assumptions made throughout the entire cost assessment process which deserve to be elaborated on beforehand.

First, it is important to realize that all of the various field instrumentation used for salinity survey work actually measure soil conductivity, *and not soil salinity per se*. Furthermore, the conversion from conductivity to salinity requires either (1) the accurate measurement or estimation of additional soil properties at each and every conductivity survey site, or (2) the collection of a limited set of "calibration" soil samples from the survey area under study. The latter approach is usually more cost effective, since acquiring soil samples at a few survey sites tends to be cheaper than acquiring soil property information at every survey site. (This approach is often referred to as "stochastic calibration", since it is based on statistical and/or geostatistical modelling techniques.) Therefore, throughout this discussion it is assumed that all but one of the survey instrumentation techniques require the collection of at least a few calibration soil samples, in addition to the collection of the survey data. The one exception is the insertion four-electrode survey, since this type of survey directly facilitates the estimation of the above mentioned secondary soil properties at each and every survey site.

Second, there are two methods which one could use to describe the average cost of a survey instrument over its expected life time. The first would be on a cost per reading basis, and the second would be on a cost per time basis. Most of the instruments typically used for field conductivity surveys (such as the Geonics EM38 or Martek SCT-10 meter) go immediately into

TABLE 12
Addresses for salinity instrumentation companies

Product	Company
Mechanized Salinity Assessment Vehicle	**Agricultural Industrial Manufacturing, Inc**. P. O. Box 53 Lodi, California 95241 USA; Telephone: 209-369-1994
	Veris Technologies 601 N. Broadway Salina, Kansas USA 67401 Telephone: 913-825-1978 Fax: 913-825-2097 web: www.veristech.com
Custom GPS / EM38 Integration and / or Software Interfacing	**Geomar Geophysics, Ltd; Attn: Jerzy Pawlowski** 2-3415 Dixie Road, Suite 348; Mississauga, Ontario Canada L4Y 4J6 Telephone: 905-306-9215; fax: 905-276-8158 e-mail:jerzy@geomar.com
EM38	**Geonics Limited** 1745 Meyerside Drive, Unit 8 Mississauga, Ontario Canada L5T 1C6 Telephone: 905-670-9580 fax: 905-670-9204 web: www.geonics.com
Conductivity Meter (CO150), Soil Cup System and Salinity and Sodicity Kits	**Hach Company** P. O. Box 389 Loveland, Colorado USA 80539 Telephone: 800-227-4224 (in USA only), otherwise 970-669-3050 fax: 970-669-2932 web: www.hach.com
Four electrode sensor, SCT-10	**Martek Instruments, Inc.** 2609 Discovery Drive, Suite 125 Raleigh, North Carolina 27616, USA Telephone: 800-628-8834 web:www.4martek.com
	Eijkelkamp Agrisearch Equipment P O Box 4 6987 Z G Giesbeek, The Netherlands Telephone: + 31 313 631941 Fax: + 31 313 632167 web: www.eijkelkamp.com
	Elico Limited B-17, Sanathnagar Industrial Estate Hyderabad, 500 018, India Telephone: 040-22-2221 Fax: 040-31-9840
Soil Temperature Equipment	**Wahl Instruments, Inc.** 5750 Hannum Avenue Culver City, California USA Telephone: 310-641-6931 Fax: 310-670-4408
TDR Equipment	**Environmental Sensors Inc.** 100-4243 Glanford Avenue Victoria, BC, Canada V8Z 4B9 Telephone: 250-479-6588 Fax: 250-479-1412

a "scan" mode the moment they are turned on. Thus, these meters are continuously analysing signal data and calculating conductivity readings, regardless of whether the surveyor actually records a reading once per second or once per hour. Therefore, it seems appropriate to use the cost per time unit method for estimating survey equipment expenses.

Operational Costs Associated with Conventional Sampling

In order to facilitate a reasonable cost analysis of the various instrumental approaches, it is first necessary to establish the overall cost associated with a conventional sampling approach. In this context, "conventional sampling" means that all salinity information is acquired through the direct laboratory analysis of soil samples, without benefit of any type of secondary instrument information.

Table 13 documents the various costs associated with such an approach for a typical 64-hectare field. The following assumptions have been made in the analysis shown in Table 13. First, the sampling is performed on a 12 by 12 grid, yielding 144 sample sites, and each site is sampled at 3 depths (0-30 cm, 30-60 cm, and 60-90 cm). Second, this process (the collection of three samples at one site) takes a two-person crew 15 minutes to complete, and this crew can walk across the field at a rate of 4 km per hour and lay out the survey grid as they walk. Third, the total walking distance is equal to 2 plus the number of transects times the physical length of the transect; note that the factoring in of 2 extra transects is done to cover the additional travel distance in between transects and the return distance (back to the starting point) after leaving the last sample site. Fourth, the total cost associated with each sample on a per sample basis is US$20.00, which includes the labelling and packaging in the field (US$0.50), handling and shipping costs (US$1.50), and external laboratory analysis costs (US$18.00). Fifth, after the soil salinity measurements are returned from the external laboratory, it will take 4 hours to input and process this data and generate a field salinity map. And finally, the two-person sampling crew is comprised of two technical support personnel paid at a rate of US$8.00 each per hour, and the data processing is performed by one technical specialist paid at a rate of US$20/hour.

TABLE 13
Cost analysis for a salinity appraisal of a typical 64-hectare field using a conventional soil sampling approach

Assumptions	800 m by 800 m field size (64 ha); 12 by 12 grid (144 sample sites); 3 sample depths per site (0-30, 30-60, 60-90 cm depths)
Time:	Sampling : 5 minutes per 30 cm sample increment; Walking speed : 4 km/hour
Sampling costs:	Labelling / packaging US$0.50/sample Shipping / handling US$1.50/sample Laboratory analysis US$18/sample Total cost US$20/sample
Total walking distance	(number of transects +2) x (transect length) = 14 x (0.8 km) = 11.2 km
Total walking time	11.2 / 4.0 = 2.8 hours
Total sampling time	(5 minutes / sample) x (3 sample depths) x (144 sites) / 60 = 36 hours
Data processing time	4 hours
Labour costs	Specialist at US$20/hour Technical support at US$8/hour
Crew size	2 Tech Support for Sampling, 1 Specialist for data processing
Total labour costs	38.8 hours @ US$16/hour + 4 hours @ US$20/hour =US$700.80
Total sampling costs	(144 cores) x (3 samples / core) @ US$20/sample = US$8 640
Overall Survey Cost	**US$9 340.80 per field (64-hectare) or US$145.95/ha**

The total labour costs associated with this survey would be US$700.80 and the total sampling costs would be US$8 640. Hence, the overall survey cost associated with the conventional sampling approach would be US$9 340.80 per field (64-hectare) or US$145.95/ha.

From a farming for profit perspective, an operational cost of approximately US$146.00/ha is inordinately expensive, and hence non-justifiable. Furthermore, as Table 13 shows, the vast majority of this cost is incurred from the commercial laboratory analysis of so many soil samples. Therefore, any techniques which can be used to reduce the total soil analysis cost will also clearly generate the greatest financial savings (with respect to the overall survey cost).

Obviously, one way to lower the analytical costs is to simply collect less samples. Of course, this is the primary idea behind each of the various instrument appraisal methods; i.e., to use the instrument readings as surrogate information (in place of actual soil samples). However, another way to achieve significant savings in the overall analytical cost is to forego the use of an commercial laboratory, and instead use some practical form of internal salinity appraisal method. In other words, one would set up their own laboratory and perform their own salinity analyses.

The simplest (and most cost effective) way to run an internal "laboratory" for soil salinity determination is to use the saturated paste salinity appraisal methods described in Rhoades, 1989b,c. In practice, such a laboratory would actually be nothing more than a work area having a sink, portable balance, a few basics laboratory supplies, and a conductivity instrument capable of measuring the conductivity of a saturated soil paste.

Table 14 documents the cost savings which can be achieved by employing such an approach in conjunction with the Hach CO150 Conductivity System. The following assumptions have been made in the analysis shown in Table 14. First, the cost of the CO150 system is assumed to be US$1,500 and its expected lifespan is 15,000 samples, yielding an equipment cost of US$0.10 per sample. Second, the average time to make a saturated soil paste is assumed to be 6 minutes, the average time to operate the meter and measure the paste is assumed to be 2 minutes, and that all other additional laboratory costs incurred during this process amount to US$0.25 per sample, and third, one technical support person is used to make the soil paste, and one technical specialist is employed to operate the conductivity meter. Using these assumptions, the total labour cost incurred during the measurement of one sample would be US$1.46 and the total equipment cost would be US$0.35. This implies that the total salinity appraisal cost would be US$2.31 per sample (after adding in the US$0.50 per sample field labelling and packaging costs), which represents a 88% reduction in the laboratory analysis cost (down from US$20/sample). Hence, the recalculated cost of the conventional sampling method becomes US$1 698.72 per field (64-hectare) or US$26.54/ha.

The cost savings calculated above represent the typical savings one would achieve by using the saturated soil paste appraisal method. Furthermore, the external laboratory reference cost of US$18/sample is conservative compared to most current US commercial laboratory rates (and hence the actual savings could be greater). None the less, these figures demonstrate that the CO150 conductivity system would pay for itself after only one field survey.

TABLE 14
Cost analysis associated with the saturated paste salinity appraisal method.

Cost of Hach CO150 System	US$1 500
Expected life span	15 000 samples
Cost per sample	US$0.10 / sample
Other laboratory costs	US$0.25/sample
Average time to make saturated paste	6 minutes
Average time to measure saturated paste	2 minutes
Laboratory crew size	1 Technical support for making paste, 1 Specialist to operate equipment
Labour cost per sample	0.10 hours/sample @ US$8.00/hour + 0.033 hours/sample @ US$20/hour = US$1.46/sample
Lab cost per sample	US$0.10 per sample (CO150) + US$0.25 per sample (supplies)
Revised sampling costs:	Labelling/packaging US$0.50/sample Shipping/handling eliminated Laboratory analysis US$1.81/sample Total cost US$2.31/sample
Recalculated Cost of Conventional Sampling Method	
Total labour costs	38.8 hours @ US$16/hour + 4 hours @ US$20/hour = US$700.80
Total sampling costs	(144 cores) x (3 samples/core) @ US$2.31/sample US$997.92
Revised Overall Survey Cost	**US$1 698.72 per field (64-hectare) or US$26.54/ha**

In general, significant cost savings can be realized using the saturated soil paste appraisal method. Furthermore, when this approach is used in conjunction with the various instrumentation methods, it becomes possible to achieve a substantial reduction in the total cost of a typical salinity survey. Therefore, although the remainder of this document will focus primarily on the various instrument appraisal costs, one should keep in mind that the greatest cost savings are always achieved through a combination of these two approaches.

Operational Costs Associated with Survey Instrumentation

As previously stated, the basic idea behind the various instrument appraisal methods is to exploit the instrument readings as surrogate information in place of actual soil samples. In principle, soil salinity can be calculated from soil conductivity provided the following additional soil information is known (or accurately estimated): temperature, saturation percentage, volumetric soil water content, and bulk density (Rhoades, 1989a). However, this additional soil information is usually not acquired during most surveys (with the exception of insertion four-electrode surveys). Hence, it usually becomes necessary to acquire a limited set of additional soil samples. These soil samples are typically referred to as "calibration" samples, because they are used to calibrate the soil conductivity to soil salinity through various statistical modelling techniques (such as regression or geostatistical models, etc.).

A formal review of the various stochastic calibration techniques is beyond the scope of this discussion (the interested reader should refer to the main text of this book or Lesch *et al.*, 1995a,b). However, the following brief comments are in order. First, an insertion four-electrode conductivity reading can generally be used to supply a more accurate estimate of soil salinity then a non-invasive (EM or horizontal array) reading. This is true because (1) the insertion four-electrode reading is depth specific, and (2) to acquire this reading one must first bore a hole into the soil, which in turn means that the soil removed from the bore hole is available for physical inspection (and hence the above mentioned secondary soil properties can be inferred from this inspection). Second, in most field survey applications an ordinary regression model can be used for purposes of calibration. And third, any type of instrument survey will generally require calibration soil samples in order to achieve maximum prediction accuracy when accurate knowledge of the secondary soil physical properties is unavailable. Therefore, in the analysis which follows, the instrumental salinity survey costs have been broken down into two separate components: (1) the costs associated with acquiring the actual instrument survey data, and (2) the costs associated with acquiring the calibration soil samples.

Table 15 documents the costs associated with acquiring EM38 survey data in a typical 64 hectare field. The following assumptions have been made in the analysis shown in Table 15. First, the EM38 survey is performed on a 12 by 12 grid (yielding 144 survey sites), two EM38 readings are acquired each site, and the data acquisition at each site takes 15 seconds. Second, the total walking distance and walking time is assumed to be the same as the distance and time required in the conventional sampling approach. Third, one hour is needed to perform all the post-survey EM38 data processing, and both the survey and data processing can be performed by one technical specialist (at a rate US$20/hour), and fourth, the total equipment cost is US$10 970, the expected equipment lifespan is 4 000 hours, and hence the average equipment cost can be estimated to be US$2.74 per hour.

The total labour costs associated with this survey would be US$88 and the total equipment costs would be US$9.32. Hence, the overall cost associated with the EM38 survey would be US$97.32 per field (64-hectare) or US$1.52/ha.

TABLE 15
Operational survey costs associated with an EM38 survey in a typical 64-hectare field

Assumptions:	800 m by 800 m field size (64 ha); 12 by 12 grid (144 survey sites); 2 survey readings per site (horizontal and vertical)
Equipment cost:	US$7 195 (EM38) + US$3 775 (DL720) = US$10 970
Expected life span:	4 000 hours
Cost per hour:	US$10 970/4 000 = US$2.74/hour
Time:	EM38 survey readings (15 seconds/site); Walking speed: 4 km / hour)
Total walking distance	(number of transects +2) x (transect length) 14 x (0.8 km) =11.2 km
Total walking time	11.2 / 4.0 = 2.8 hours
Total survey time	(0.25 min) x (144 sites) / 60 min = 0.6 hours
Data processing time	1 hour
Labour costs:	Specialist US$20/hour
Crew size	1 Specialist for EM38 survey work and data processing
Total labour costs	4.4 hours @ US$20/hour = US$88
Total equipment costs	3.4 hours @ US$2.74 per hour = US$9.32
Overall survey cost	**US$97.32 per field (64-hectare) or US$1.52/ha**

Table 16 documents the costs associated with acquiring SCT-10 insertion four-electrode survey data in the same 64-hectare field. The following assumptions have been made in the analysis shown in Table 16. First, the insertion four-electrode survey is performed on a 12 by 12 grid (yielding 144 survey sites), three insertion four-electrodes are acquired each site (at depths of 15, 45, and 75 cm), estimates of the secondary soil physical properties are acquired during this process, and all of the data acquisition at each site takes 4.5 minutes. Second, the total walking distance and walking time is assumed to be the same as the distance and time required in the conventional sampling approach. Third, one hour is needed to perform all the post-survey SCT-10 data processing, and both the survey and data processing can be performed by one technical specialist (at a rate US$20/hour). And forth, the total equipment cost is

US$2 500, the expected equipment lifespan is 4,000 hours, and hence the average equipment cost can be estimated to be US$0.63/hour.

TABLE 16
Operational survey costs associated with an insertion four electrode survey (using a Martek SCT-10 meter) in a typical 64-hectare field

Assumptions	800 m by 800 m field size (64 ha); 12 by 12 grid (144 survey sites) 3 survey readings per site (0.15, 0.45, and 0.75 cm depths)
Equipment cost	US$2 000 (SCT-10 meter) + US$500 (Probe) = US$2 500
Expected life span	4 000 hours
Cost per hour	US$2 500/4,000 hours = US$0.63 per hour
Time	SCT-10 Insertion 4-probe survey readings (1.5 min per 30 cm depth increment) Walking (4 km / hour)
Total walking distance	(number of transects +2) x (transect length) = 14 x (0.8 km) = 11.2 km
Total walking time	11.2 km/(4 hours/km) = 2.8 hours
Total survey time	(1.5 min) x (3 readings per site) x (144 sites)/60 min = 10.8 hours
Data processing time	1 hour
Labour costs	Specialist at US$20/hour
Crew size	1 Specialist for SCT-10 survey work and data processing
Total labour costs	14.6 hours @ US$20/hour = US$292
Total equipment costs	13.6 hours @ US$0.63 per hour = US$8.57
Overall survey cost	**US$300.57 per field (64-hectare) or US$4.70/ha**

The total labour costs associated with this survey would be US$292.00 and the total equipment costs would be US$8.57. Hence, the overall cost associated with the SCT-10 insertion four-electrode survey would be US$300.57 per field (64-hectare) or US$4.70/ha.

Table 17
Operational survey costs associated with a fixed, horizontal array survey (using a Martek SCT-10 meter) in a typical 64-hectare field

Assumptions	800 m by 800 m field size (64 ha); 12 by 12 grid (144 survey sites) 2 survey readings per site (1 m and 2 m span)
Equipment cost	US$2 000 (SCT-10 meter) + US$300 (Probe) = US$2 300.00
Expected life span	4 000 hours
Cost per hour	US$2 300/4, 000hours = US$0.58 per hour
Time	SCT-10 Horizontal Array Survey Readings (1 minute/site); Walking speed: 4 km/hour)
Total walking distance	(number of transects +2) x (transect length) = 14 x (0.8 km) = 11.2 km
Total walking time	11.2 / 4.0 = 2.8 hours
Total survey time	(1 min per site) x (144 sites)/60 min = 2.4 hours
Data processing time	1 hour
Labour costs	Specialist at US$20/hour
Crew size	1 Specialist for SCT-10 survey work and data processing
Total labour costs	6.2 hours @ US$20/hour = US$124
Total equipment costs	5.2 hours @ US$0.58 per hour = US$3.02
Overall survey cost	**US$127.02 per field (64-hectare) or US$1.98/ha**

Table 17 documents the costs associated with acquiring SCT-10 horizontal array survey data in a typical 64-hectare field. The following assumptions have been made in the analysis shown in Table 17. First, the horizontal array survey is performed on a 12 by 12 grid (yielding 144 survey sites), two array readings are acquired each site (a 1 metre and 2 metre span), and the data acquisition at each site takes 1 minute. Second, the total walking distance and walking time is again assumed to be the same as the distance and time required in the conventional sampling approach. Third, one hour is needed to perform all the post-survey SCT-10 data processing, and both the survey and data processing can be performed by one technical specialist (at a rate US$20/hour), and forth, the total equipment cost is US$2 300, the expected equipment lifespan is 4 000 hours, and hence the average equipment cost can be estimated to be US$0.58/hour.

The total labour costs associated with this survey would be US$124 and the total equipment costs would be US$3.02. Hence, the overall cost associated with the SCT-10 horizontal array survey would be US$127.02 per field (64-hectare) or US$1.98/ha.

The EM38 and SCT-10 equipment life spans have been assumed to be 4 000 hours (Tables 15, 16 and 17). These lifespan estimates are conservative, since they imply that either instrument would only last about 100 weeks (i.e., two years) when used 8 hours per day, five days a week. In reality, both instruments will last much longer than this, provided they are not seriously abused during the field survey work.

Note that the above overall cost estimates only reflect the costs associated with the instrument survey process; the costs associated with the collection of additional calibration soil samples still must be determined for two of the survey techniques. For calibration purposes, assume that the regression models associated with either the EM38 or SCT-10 horizontal array data will be estimated using 12 soil samples. Furthermore, note that one must estimate a unique regression model for each sample depth (i.e., 3 separate regression models for the three sample depths). Thus, soil samples from 12 separate sample sites must be acquired to calibrate either the EM38 or horizontal array data, implying that 36 total soil samples must be analysed.

Table 18 documents the costs associated with acquiring either EM38 or SCT-10 horizontal array calibration soil samples in the 64-hectare field. The assumptions made in Table 18 are identical to those made in Table 13, except for the following adjustments: (1) only 12 sites need to be sampled, and (2) the post laboratory data processing time can be done in one hour. Hence, the revised total labour costs for this calibration sampling become US$112.80, and the revised total sampling costs become either US$720 (if the samples are sent to an commercial laboratory) or US$83.16 (if the salinity is measured using the saturated soil paste methodology).

TABLE 18
Operational costs for the calibration soil sampling associated with either the EM38 or SCT-10 horizontal array survey

Assumptions	800 m by 800 m field size (64 ha); (12 sample sites) 3 sample depths per site (0-30, 30-60, 60-90 cm depths)
Time	Sampling (15 minutes/site); Walking speed: 4 km/hour.
Sampling costs	External Laboratory US$20/sample Internal Laboratory US$2.31 / sample
Total walking distance	(number of transects +2) x (transect length) = 14 x (0.8 km) = 11.2 km
Total walking time	11.2/4 = 2.8 hours
Total sampling time	(15 min) x (12 cores)/60 min = 3 hours
Data processing time	1 hour
Labour costs	Specialist at US$20/hour and Technical support at US$8/hour
Crew size	2 technical support for sampling and 1 specialist for data processing
Total labour costs	5.8 hours @ US$16/hour + 1.0 hours @ US$20/hour = US$112.80
Total sampling costs	(12 sites) x (3 samples / site) @ US$20/sample = US$720 (using commercial laboratory) or (12 sites) x (3 samples / site) @ US$2.31/sample = US$83.16 (using saturated paste method, internal laboratory)
Overall survey cost	**US$832.80 per field (64-hectare) or US$13.01/ha (commercial lab)** **US$195.96 per field (64-hectare) or US$3.06/ha (internal lab)**

Given these figures, it is now possible to estimate the total (composite) survey costs associated with each instrument. Table 19 lists these estimates, which represent the sum total of three separate costs: the instrument survey costs, the calibration sampling costs, and the final data processing costs. (In Table 19, the data processing time is estimated to be 6 hours regardless of which instrumentation approach is employed. This time includes performing all the stochastic or deterministic salinity calibrations and generating the final field salinity estimates and/or maps.) Note that the overall, total costs for the EM38, SCT-10 horizontal array (internal Lab), and SCT-10 insertion four-electrode surveys come to US$413.28, US$442.98, and US$420.57 per field (64-hectare), respectively (assuming internal laboratory analysis of all calibration soil samples in the first two surveys, and no collection of calibration samples in the third survey).

This last cost analysis highlights two points. First, on a total expense basis, there is relatively little difference between the three types of instrumentation survey techniques. This is because the various cost factors (i.e., survey costs, hourly instrument costs, and calibration sampling costs) tend to balance each other out. More importantly, these final figures show the tremendous cost savings which can be achieved through the judicious use of field instrumental methods in conjunction with internal laboratory (saturated soil paste appraisal) methods. Recall that the original cost for performing a conventional soil survey within this hypothetical 64-hectare field was estimated to be US$9 340.80. By employing the EM38 or either one of the SCT-10 surveys described above, one could now reduce this cost to about US$415 - US$440. In other words, one can achieve a 96% reduction in the overall survey cost.

TABLE 19
Total survey and calibration costs for the different survey instrumentation approaches

Base : Data processing time is 6 hours @ US$20/ hour for either type of survey (includes basic statistical or deterministic calibration and map generation)		
Item		US$
1. SCT-10 Survey Cost, insertion four-electrode (IF), **(Table 16)**		300.57
2. SCT-10 Survey Cost, horizontal array (HA), **(Table 17)**		127.02
3. EM38 Survey Cost, **(Table 15)**		97.32
4. Calibration Sampling Costs EM38 or SCT-10(HA)/Commercial Lab, **(Table 18)**		832.80
5. Calibration Sampling Costs EM38 or SCT-10(HA)/Internal Lab, **(Table 18)**		195.96
6. Calibration Sampling Costs SCT-10(IF)/ no soil samples		0.00
7. Data processing cost		120.00
Total survey costs = instrument survey cost + calibration sampling cost + data processing cost	US$ per field (64-hectare)	US$/ha
EM38 / Commercial Lab (Per field = Line 3 + Line 7 + US$ 832.80)	1 050.12	16.41
EM38 / Internal Lab (Per field = Line 3 + Line 7 + US$ 195.96)	413.28	6.46
SCT-10(HA) / Commercial Lab (Per field = Line 2 + Line 7 + US$ 832.80)	1 079.82	16.87
SCT-10(HA) / Internal Lab (Per field = Line 2 + Line 7 + US$ 195.96)	442.98	6.92
SCT-10(IF) / no soil samples (Per field = Line 1 + Line 7)	420.57	6.57

Conventional Sampling versus Survey Instrumentation Costs in Multi-Field (Large Area) Survey Applications

The previous cost analysis discussion has been based on a comparison of costs for surveying a single field. In practice, a large scale salinity survey will typically encompass many fields (and/or a fairly large survey area). Therefore, a comparison of the conventional sampling versus survey instrumentation costs inherent in such a large scale survey is obviously desirable.

Table 20 documents the conventional sampling and survey instrumentation costs associated with a large scale salinity survey encompassing 6 400 hectares of continuous agricultural land. The following assumptions have been made in the analysis shown in Table 20. First, the 6 400-hectare area is assumed to be comprised of 100 individual 64-hectare fields, and each field is to be sampled (or surveyed) on a 12 by 12 grid. Second, for the conventional sampling, all the time and cost assumptions stated in Table 13 are assumed to be the same. Third, for the survey instrumentation process, each field is surveyed using an EM38 meter. Hence, all the time and instrument survey cost assumptions stated in Table 15 are assumed to be the same. Additionally, it is assumed that 4 sites from each field are selected for calibration soil sampling (generating 400 sample sites and hence 1 200 soil samples across the survey area), and that this calibration sampling cost can be estimated to cost US$72.92 on a per field basis (see Table 21). Third, an additional cost of US$3 000 (for purchasing a single GPS unit) is assumed to be incurred by either survey process. Fourth, for the survey instrumentation process, an additional cost of US$4 000 is incurred for 80 hours worth of geostatistical analysis (for purposes of conductivity-to-salinity calibration and map generation). And finally, the conventional sampling approach uses a commercial laboratory to perform all of the laboratory salinity analyses, and the calibration soil samples are analysed for salinity using the saturation paste conductivity method (i.e., internal laboratory).

TABLE 20
Cost comparison between the conventional versus survey instrumentation and calibration methodologies for a 6 400-hectare survey (one hundred 64-ha fields).

Assumptions: (i)100 fields, each 800 m by 800 m, all located across one continuous area, (ii) 12 by 12 grid (144 sites) per field,(iii) Conventional approach requires 144 sample sites (3 samples per depth), (iv) Composite approach requires 144 survey sites (2 readings per site) + 4 sample sites for calibration (3 sample per depth), (v) Both approaches require the purchase of 1 GPS unit @ US$3,000 per unit, (vi) Conventional approach uses commercial laboratory and (vii) Composite approach uses internal laboratory (saturated paste method)	
Total conventional survey cost	= 100 x (US$9 340.80) + GPS cost = US$934 080 + 3,000 = US$937 080
Total instrument survey cost	100 x (US$97.32) = US$9,732
Total calibration sampling cost	100 x (US$72.92) = US$7,292
GPS cost	US$3,000
Total data analysis cost (for geostatistical analysis)	80 hours @ US$50/hour = US$4 000
Total instrument Survey/Calibration/Analysis Cost	US$24 024

Notes: (i) Overall time to complete conventional survey is 4,280 hours, not including external laboratory time for salinity analysis and (ii) Overall time to complete instrument survey / calibration analysis is 950 hours, including internal laboratory salinity analysis and contracted data analysis.

The overall cost of the conventional sampling approach would be US$937 080 (US$146.42/ha), while the overall cost of the survey instrumentation approach would be only US$24 024 (US$3.75/ha). Hence, by exploiting the use of the survey instrumentation approach, the composite cost of the entire survey process would be reduced by over 97%.

Some details inherent to the assumptions in Table20 are worth expanding on. It has been assumed that EM38 instrumentation is being used in this analysis, because this survey process is the fastest (and hence the least expensive). However, for calibration purposes, only four samples per field are acquired. The reason why this further reduction in the number of soil samples is possible is due to the fact that the 6 400-hectare survey area is assumed to be continuous. Hence, because the survey has been conducted across one large area, it will be possible (and in fact, preferable) to employ more sophisticated geostatistical techniques to estimate the calibration equation(s). Note that such an analysis would have to typically be performed by a trained statistician, implying an additional data analysis expense (in this example, US$4 000). However, this additional expense is more than offset by the cost savings achieved through the reduction of the soil sampling requirements.

Second, the total survey time under the survey instrumentation approach is significantly faster than the total time under the conventional sampling approach. The instrumentation approach requires a total of 950 hours; which represents the sum of 340 EM38 survey hours, 220 soil sampling hours, 150 data processing hours, 160 laboratory analysis hours, and 80 hour of external data analysis. In contrast, the conventional sampling approach would require a total of 4 280 hours, not including the time required for the commercial laboratory analysis of the soil samples. Additionally, since 340 hours < 4 000 (equipment life span), all of the EM38 survey work can be done with one EM38 and since 1 200 (number of calibration samples) < 15 000 (equipment life span in sample number) all of the saturated paste soil conductivity measurements can be made with one CO150 conductivity meter.

Finally, it should be pointed out that this sort of non-invasive survey / calibration sampling process is not necessarily the most cost effective approach if the survey region is highly discontinuous and/or the survey grid is extremely large. If 100 separate, discontinuous fields are to be surveyed, then in general no reduction in the per field calibration sampling size is possible. Hence, the cost differences between the three survey instrumentation techniques are rather minimal (see Table 19). Likewise, suppose a very coarse survey grid was to be employed over a large area; for example, a 1 km by 1 km grid over a 10 000-km^2 area. Then the variation in the secondary soil physical properties would undoubtedly be quite large and the site to site correlation between these properties would probably be minimal. Hence, the insertion four-electrode technique would be expected to yield the most accurate information for the least cost, since the calibration sampling requirements for the other two techniques would be greatly increased.

TABLE 21
Operational costs for the calibration soil sampling associated with 6 400-hectare survey discussed in Table 20

Assumptions	800 m by 800 m field size (64 ha); (4 sample sites per field) 3 sample depths per site (0-30, 30-60, 60-90 cm depths) and internal laboratory (saturated paste) appraisal methods.
Time	Sampling (15 minutes/site); Walking speed: 4 km/hour.
Sampling costs	Internal Laboratory US$2.31 / sample
Total walking distance	(number of transects +2) x (transect length) 6 x (0.8 km) = 4.8 km
Total walking time	4.8 / 4.0 = 1.2 hours
Total sampling time	(15 minutes/site) x (4 sites) / 60 min = 1 hour
Data processing time	0.5 hour
Labour costs	Specialist at US$20/hour and Technical support at US$8/hour
Crew size	2 Technical support for sampling, 1 Specialist for data processing
Total labour costs	2.2 hours @ US$16/hour + 0.5 hours @ US$20/hour = US$45.20
Total sampling costs	(4 sites) x (3 samples / site) @ US$2.31 / sample US$27.72 (using saturated paste method, internal laboratory)
Overall survey cost	**US$72.92 per field (64-hectare) or US$1.14/ha**

Cost Advantages associated with Instrument Mobilization

As the previous example demonstrates, it is possible to survey fairly large areas without mobilizing any of the instrumentation equipment. However, as the survey regions become increasingly large, the additional cost saving advantages of instrument mobilization can become quite significant.

There are three primary cost advantages associated with the mobilization of the various salinity survey instruments. The first advantage is the speed in which the survey process can be completed. Mechanized systems can almost always be used to survey more land than hand held systems, simply because of their increased travel speed. Second, mechanized systems can be used to collect significantly more survey data. Indeed, many of the commercial systems currently available can collect survey data in a nearly continuous fashion. And third, when hand held units are mounted or adapted into mobilized platforms, they simply tend to last longer. This increased lifespan is primarily due to the significant reduction in operator handling and/or manual abuse that these units receive.

TABLE 22
Typical start up costs for three types of mobilized instrument assessment systems

Dual EM38 Trailering System	***US$***
EM38 Meters (2x US$7,195)	14 390
DL720 data logger	3 775
Trailer Platform	3 000
Instrument interfacing	4 500
GPS Unit	3 000
Alternative terrain vehicle (ATV)	6 000
Total system cost	**34 665**
Expected system Life	10,000 hours
Equipment maintenance	50%(ATV cost) + 10%(Trailer cost) US$3 300 over expected system life
Cost per hour	(34 665 + 3 300) / 10 000 = US$3.80/hour
Mobilized, Tractor Mounted Horizontal Array System	***US$***
Horizontal array system	5 250
GPS Unit	3 000
Tractor	12 500
Total system cost	**20 750**
Expected System Life	10 000 hours
Equipment Maintenance	50%(Tractor cost) US$6 250 over expected system life
Cost per hour	(20 750 + 6 250)/10 000 = US$2.70 per hour
Mobilized, Multi-Instrument Transport System	**US$**
Basic hydraulic transport vehicle	8 000
Structural fabrication costs	27 933
EM38 Meter	7 195
DL720 data logger	3 775
SCT-10 Meter	2 000
GPS Unit	3 000
Instrument interfacing	1 400
Total system cost	53 303
Expected system life	10 000 hours
Equipment Maintenance	50%(Basic Transport Vehicle cost) + 10% (Fabrication cost) = US$6 793 over expected system life
Cost per hour	(53,303 + 6,793) / 10,000 = US$6.01 per hour

Notes: secondary vehicle costs (for transporting systems from field to field) are not included in these cost per hour estimates.

Table 22 lists the approximate start up costs for three mechanized salinity assessment systems; a dual EM38 trailering system, a mobilized horizontal array system, and a mobilized, multi-instrument transport system. The first two systems can be used to rapidly acquire bulk soil conductivity information in a continuous manner, while the third system is primarily designed to rapidly collect more depth specific conductivity information in a stop and go manner. Hence, the dual EM38 trailering system and mobilized horizontal array system are ideally suited for collecting large amounts of average soil profile conductivity information. In contrast, the mobilized, multi- instrument transport system is ideally suited for rapidly acquiring multiple sets of soil conductivity data over a grid of fixed locations. Hence, this system supplies more accurate, depth specific information, which in turn makes it more useful for repetitive salinity monitoring activities and/or any large scale inventorying process with requires high vertical (soil profile) resolution.

The total start up cost for the dual EM38 trailering system includes the following: two EM38 units (one each for horizontal and vertical readings), one DL720 data logger, the trailer platform (estimated to cost US$3 000), the electronic instrument interfacing (estimated to cost US$4 500), one GPS unit, and one ATV. Assuming that the GPS unit and ATV can be purchased for US$3 000 and US$6 000, respectively, the total system cost would be US$34 665. Additional equipment maintenance expenses (for maintaining the ATV and trailer) should also be factored into the system cost; in Table 22 this cost has been estimated to be approximately US$3 300 over the life of the system. Assuming that the system life is 10 000 hours (which is roughly equivalent to 5 years, if the system is operated 8 hours a day, 5 days a week), the average cost per hour for operating this dual EM38 trailering system would be US$3.80. Note that this hourly cost does not include any fuel or labour costs associated with the surveying process.

The total start up cost for the tractor mounted, horizontal array system would be US$20 750, after factoring in the cost of the horizontal array unit, the GPS unit, and a tractor. The additional equipment maintenance cost for maintaining a tractor would probably be somewhat higher; in Table 22 it has been estimated to be approximately US$6 250over the life of this system. Thus, assuming the same sort of system life expectancy (10 000 hours), the average cost per hour for operating this system would be US$2.70. Once again, this hourly cost does not include any fuel or labour costs associated the surveying process.

In a similar manner, the total start up cost for the multi-instrument transport system would be US$53 303, after factoring in all of the various equipment costs, structural fabrication costs, and the base cost of the initial hydraulic transport vehicle. Additionally, the estimated equipment maintenance costs for this system would be US$6 793 of the expected system life. Again, assuming a system life of 10 000 hours, the average cost per hour for this system would be US$6.01 (not including fuel or labour costs).

Note that in all three start up cost examples, no additional secondary vehicle costs have been incorporated into the analysis. However, some sort of towing vehicle would typically be needed if any of these systems were to be transported over significant distances. Additionally, the start up costs for both the dual EM38 and the tractor mounted horizontal array systems could be significantly reduced, if one was already in possession of a suitable ATV or tractor.

Based on the hourly cost estimates discussed above, it is now possible to calculate the total instrument survey cost one would incur in a survey of a typically 64-hectare field using each of these three systems. These survey costs are documented in Tables 23, 24 and 25 for the dual EM38, horizontal array, and multi-instrument transport systems, respectively.

The assumptions used in the Table 23 calculations are as follows. The dual EM38 trailering system is assumed to traverse the field at a rate of 12 km per hour, and the survey is performed across 24 equally spaced 800 meter transects. Thus, the total travel distance would be 20.8 km, and the system fuel costs are assumed to be US$0.15 per km. Furthermore, the system can still be operated by one technical specialist (paid at a rate of US$20/hour), and the post survey data processing still takes one hour. Based on these assumptions, the overall mechanized EM38 survey cost for this 64-hectare field would come to US$64.29, or about US$1.00 per hectare.

TABLE 23
Operational survey costs associated with a mobilized EM38 survey in a typical 64-hectare field (dual EM38 trailering system costs)

Assumptions	800 m by 800 m field size (64 ha) 24 transects (readings collected every 1 to 10 seconds per transect) (576 to 5760 total survey sites)
Cost per hour	US$3.80
Travel time	12 km / hour
Total travel distance	(number of transects +2) x (transect length) 26 x (0.8 km) = 20.8 km
Total travel time	20.8 / 12.0 = 1.73 hours (includes survey time)
Total fuel costs	(US$0.15 per km) x 20.8 km = US$3.12
Data processing time	1 hour
Labour costs	Specialist at US$20/hour
Crew size	1 Specialist for automated survey work and data processing
Total labour costs	2.73 hours @ US$20/hour = US$54.60
Total equipment costs	1.73 hours @ US$3.80 per hour + fuel cost = US$9.69
Overall survey cost	**US$64.29 per field (64-hectare) or US$1/ha.**

A nearly identical set of assumptions in Table 24 yield an overall horizontal array survey cost of US$83.18, or about US$1.30 per hectare. Note that the only differences in the Table 24 assumptions concern the travel time and fuel costs. Direct contact horizontal array systems tend to be towed across a field at a somewhat slower rate, and a tractor would be expected to use slightly more fuel.

TABLE 24
Operational survey costs associated with a mobilized, horizontal array conductivity survey in a typical 64-hectare field

Assumptions	800 m by 800 m field size (64 ha); 24 transects (readings collected every 1 to 10 seconds per transect) and 864 to 8640 total survey sites
Cost per hour	US$2.70
Travel time	8 km/hour
Total travel distance	(number of transects +2) x (transect length) = 26 x (0.8 km) = 20.8 km
Total travel time	20.8 / 8.0 = 2.6 hours (includes survey time)
Total fuel costs	(US$0.20 per km) x 20.8 km = US$4.16
Data processing time	1 hour
Labour costs	Specialist at US$20/hour
Crew size	1 Specialist for automated survey work and data processing
Total labour costs	3.6 hours @ US$20/hour = US$72.00
Total equipment costs	2.6 hours @ US$2.70 per hour + fuel cost = US$11.18
Overall survey cost	**US$83.18 per field (64-hectare) or US$1.30/ha**

The assumptions shown in Table 25 are slightly different. A mobilized, multi-instrument transport system would generally be used to perform a grid survey similar to the earlier manual survey instrumentation processes (only faster and in more detail). Hence, in Table 25 the same 12 by 12 survey grid is used. Additionally, 4 EM38 and 4 horizontal array survey readings are acquired at each site, and the total data acquisition time is assumed to take 45 seconds (per site).

As in Table 23, the travel time is assumed to be 12 km/hour, the fuel costs are assumed to be US$0.15/km, one technical specialist can still operate the system, and data processing still takes one hour. Based on these assumptions, the overall multi-instrument transport system survey cost for this 64-hectare field would come to US$92.69, or about US$1.45/ha.

TABLE 25
Operational survey costs associated with a mobilized, multi-instrument survey in a typical 64-hectare field (multi-instrument transport system costs)

Assumptions	800 m by 800 m field size (64 ha); 12 by 12 grid (144 survey sites) and 4 EM38 and horizontal array survey readings per site.
Cost per hour	US$6.01/hour
Travel time	12 km/hour
Survey time	45 seconds per site
Total travel distance	(number of transects +2) x (transect length) = 14 x (0.8 km) = 11.2 km
Total travel time	11.2/12.0 = 0.93 hours
Total survey time	(0.75 min) x (144 sites)/60 min = 1.8 hours
Total fuel costs	(US$0.15 per km) x 11.2 km = US$1.68
Data processing time	1 hour
Labour costs	Specialist at US$20/hour
Crew size	1 Specialist for automated survey work and data processing
Total labour costs	3.73 hours @ US$20/hour = US$74.60
Total equipment costs	2.73 hours @ US$6.01 per hour + fuel cost = US$18.09
Overall survey cost	**US$92.69 per field (64-hectare) or US$1.45/ha**

In all three cases, the final per hectare costs can be seen to be lower than the lowest manual instrument survey cost, which was US$1.52 per hectare for the EM38 (Table 15). Furthermore, the amount of instrument survey data collected during each of the automated survey processes has been greatly increased. For example, given the assumptions in Table 23, the dual EM38 trailering system would complete each survey transect in 4 minutes. If one set of readings (horizontal and vertical) are acquired every 5 seconds, then this automated system can effectively collect signal data at 1152 survey sites within the field (24x240/5=1 152). Hence, either the dual EM38 or tractor mounted horizontal array systems can be used to greatly increase the spatial resolution of the conductivity data without raising the per hectare survey instrumentation costs. Likewise, the multi-instrument transport system can effectively collect 4 to 8 times the amount of conductivity data in the same amount of time over each survey site, compared to manual survey methods. Thus, the vertical resolution of the survey process is greatly increased, again without raising the per hectare survey instrumentation costs.

For very large survey applications, instrument mobilization becomes highly cost effective. For example, each of these mobilized systems could be used to effectively survey thousands of hectares per week in typical, large scale agricultural regions. Without question, this represents a far greater amount of landscape than can ever be effectively inventoried in the same amount of time using manual survey techniques.

CONCLUSION

A comprehensive cost analysis of the various survey instrumentation techniques currently used for soil salinity assessment has been carried out. It included a description of the capital costs associated with both manual and mobilized instrumentation techniques, useful analytical laboratory techniques, and current software. Additionally, a detailed analysis of the operational costs incurred when applying these various techniques for the measurement of soil salinity has been performed.

Table 26 summarizes the operational costs associated with the various survey instrumentation techniques discussed herewith. The operational cost of the three most common manual instrumentation techniques were all found to be between US$6.40 to US$6.90 per hectare in a typical 64-ha field. These operational cost figures were comprised of three components; the actual survey instrumentation costs, the data analysis costs, and the calibration sampling costs (when necessary). In all three cases, these figures were about 96% less than the conventional soil sampling costs, which were estimated to be about US$146/ha. Additionally, the cost analysis of the three most common mechanized instrumentation systems demonstrated that these per hectare costs can be further reduced by suitably mobilizing the various instruments, especially when large agricultural regions are to be surveyed.

Overall, this cost benefit analysis suggests that the proper implementation of the various survey instrumentation techniques can lead to significant financial savings in most all types of salinity survey applications. Furthermore, these savings would be expected to be greatest in large scale agricultural surveys and/or regional inventorying applications.

TABLE 26
Summary table of various survey instrumentation costs discussed in Tables 13 through 25. All cost and time factors based on a 64-hectare field size, and time factors do not include commercial or internal laboratory analysis time components

Survey Method	Cost, US$ /ha	Time hours/field	Number of Survey Sites per field
Individual 64-ha Field			
Conventional sampling (1)	145.95	42.8	144
Conventional sampling (2)	26.54	42.8	144
EM38 manual (3)	6.46	4.4	144
SCT-10 HA manual (3)	6.92	6.2	144
SCT-10 IF manual (3)	6.57	14.6	144
EM38 mobilized (4)	5.94	2.7	576+
SCT-10 HA mobilized (4)	6.24	3.6	576+
Multi-instrument mobilized (4)	6.39	3.7	144
Continuous 6400 ha Region:			
EM38 mobilized (5)	3.24	2.7	576+
SCT-10 HA mobilized (5)	3.53	3.6	576+
Multi-instrument mobilized (5)	3.68	3.7	144

Notes: (1) from Table 13; (2) from Table 14; (3) from Table 19; (4) from Tables 23-25 & 19 (substituting mobilized for manual survey costs) and (5) from Tables 23-25 & 20 (substituting mobilized for manual survey costs).

References

Adams, F. 1974. Soil solution. *In* E.W. Carson (ed.) The plant root and its environment. Univ. Press of Virginia, Charlottesville, p. 441-481.

Adams, F., C. Burmester, N. V. Hue, and F. L. Long. 1980. A comparison of column -displacement and centrifuge methods for obtaining soil solutions. Soil Sci. Soc.Am. J. 44:733-735.

Adiku, S. G. K., M. Renger, and C. Roth. 1992. A simple model for extrapolating the electrical conductivity data of gypsum containing soils from reference soil extract data. Agri. Water Mgt. 21:235-246.

Allison, L.E. 1973. Oversaturation method for preparing saturation extracts for salinity appraisal. Soil Sci. 116:65-69.

Austin, R. S., and J. D. Oster. 1973. An oscillator circuit for automated salinity sensor measurements. Soil Sci. Soc. Am. Proc. 37:327-329.

Austin, R.S., and J. D. Rhoades. 1979. A compact, low-cost circuit for reading four-electrode salinity sensors. Soil Sci. Soc. Am. J. 43:808-810.

Barnes, H. E. 1954. Electrical subsurface exploration simplified. Roads and Streets, May 1954, pp. 81-84,140.

Beatty, H.J., and J. Loveday. 1974. Soluble cations and anions. *In* J. Loveday (ed.) Methods for analysis of irrigated soils. Technical Communication No. 54, Commonwealth Bureau of Soils, Commonwealth Agricultural Bureau, Farmham Royal, Bucks, England, p. 108-117.

Biggar, J. W., and D. R. Nielsen. 1976. Spatial variability of the leaching characteristics of a field soil. Water Resour. Res. 12:78-84.

Bohn, H.L., J. Ben–Asher, H.S. Tabbara, and M. Marwan. 1982. Theories and tests of electrical conductivity in soils. Soil Sci. Soc. Am. J. 46:1143—1146.

Bottraud, Jean-Christophe, and J. D. Rhoades. 1985a. Effect of exchangeable sodium on soil electrical conductivity-salinity calibrations. Soil Sci. Soc. Am. J. 49:1110-1113.

Bottraud, Jean-Christophe, and J. D. Rhoades. 1985b. Referencing water content effects on soil electrical conductivity - salinity calibrations. Soil Sci. Soc. Am. J.49:1579-1581.

Briggs, L.J., and A.G. McCall. 1904. An artificial root for inducing capillary movement of soil moisture. Sci. 20:566-569.

Cameron, D. R., E. De Jong, D.W.L. Read, and M. Oosterveld. 1981. Mapping salinity using resistivity and electromagnetic inductive techniques. Can. J. Soil Sci. 61:67-78. Cameron, M. E., R. C. McKenzie and G. Lachapelle, 1994. Soil salinity mapping with electromagnetic induction and satellite-based navigation methods. Can. J. Soil Sci. 74:335-343.

Carter, L. M., J. D. Rhoades and J. H. Chesson. 1993. Mechanization of soil salinity assessment for mapping. Proc., 1993 ASAE Winter Meetings, Chicago, IL., December 12-17, 1993

Chow, T. L. 1977. A porous cup soil-water sampler with volume control. Soil Sci. 124:173-176.

Cook, P. G., M. W. Hughes, G. R. Walker and G. B. Allison. 1989. The calibration of frequency-domain electromagnetic induction meters and their possible use in recharge studies. J. Hydrology 107:251-265.

Corwin, D.L., and J. D. Rhoades. 1982. An improved technique for determining soil electrical conductivity-depth relations from above ground electromagneticm measurements. Soil Sci. Soc. Am. J. 46:517-520.

Corwin, D.L., and J. D. Rhoades. 1984. Measurement of inverted electrical conductivity profiles using electromagnetic induction. Soil Sci. Soc. Am. J. 48:288-291.

Corwin, D.L., and J. D. Rhoades. 1990. Establishing soil electrical conductivity - depth relations from electromagnetic induction measurements. Commun. Soil Sci. Plant Anal. 21:861-901.

Dao, T.H., and T. L. Lavy. 1978. Extraction of soil solution using a simple centrifugation method for pesticide adsorption - desorption studies. Soil Sci. Soc. Am. J. 42:375-377.

Davies, B.E., and R. I. Davies. 1963. A simple centrifugation method for obtaining small samples of soil solution. Nature. 198:216-217.

Diaz, L. and J. Herrero. 1992. Salinity estimates in irrigated soils using electromagnetic induction. Soil Sci. 154:151-157.

Dobrin, M. B. 1960. Introduction to Geophysical Prospecting, McGraw-Hill Book Company, 446 pp.

Doolittle, J. A., K.A. Sudduth, N. R. Kitchen, and S. J. Indorante. 1994. Estimating depths to claypans using electromagnetic induction methods. J. Soil and Water Cons. 49:572-575.

Duke, H.R., and H.R. Haise. 1973. Vacuum extractors to assess deep percolation losses and chemical constituents of soil water. Soil Sci. Soc. Amer. Proc. 37:963-964.

Elkhatib, E.A., O.L. Bennett, V.C. Baligar, and R.J. Wright. 1986. A centrifuge method for obtaining soil solution using an immiscible liquid. Soil Sci. Soc. Am. J. 50:297-299.

Elkhatib, E.A., J.L. Hern, and T.E. Staley. 1987. A rapid centrifugation method for obtaining soil solution. Soil Sci. Soc. Am. J. 51:578-583.

Enfield, C.G., and D. D. Evans. 1969. Conductivity instrumentation for *in situ* measurements of soil salinity. Soil Sci. Soc. Amer. Proc. 33:787-789.

FAO. 1988. World Agriculture Toward 2000: An FAO Study. N. Alexandratos (ed.). Bellhaven Press, London. 338 pp.

Gerlach, F. L., and A. E. Jasumbach. 1989. Global positioning systems canopy effect study. Technol. Development Center, Forest Service, U.S. Dept. Agr., Missoula, MT., 18 pp.

Ghassemi, F., A. J. Jakeman and H. A. Nix. 1995. Salinization of Land and Water Resources: Human Causes, Extent, Management and Case Studies, CAB International, 526 pp.

Gillman, G. P. 1976. A centrifuge method for obtaining soil solution. Div. Rep. No. 16. CSIRO, Div. of Soils, Townsville, Queensland, Australia.

Grieve, A.M., E. Dunford, D. Marston, R.E. Martin, and P. Slavich. Effects of waterlogging and soil salinity on irrigated agriculture in the Murray Valley: a review. *Australian Journal of Experimental Agriculture*. 1986, 26, 761-77.

Grossman, J., and P. Udluft. 1991. The extraction of soil water by the suction-cup method: A review. J. Soil Sci. 42:83-93.

Halvorson, A. D., and J. D. Rhoades. 1974. Assessing soil salinity and identifying potential saline-seep areas with field soil resistance measurements. Soil Sci. Soc. Amer. Proc. 38:576-581.

Halvorson, A. D., and J. D. Rhoades. 1976. Field mapping soil conductivity to delineate dryland saline seeps with four-electrode technique. Soil Sci. Soc. Am. J. 40:571-575.

Halvorson, A. D., J. D. Rhoades, and C.A. Ruele. 1977. Soil salinity-four-electrode conductivity relationships for soils of the Northern Great Plains. Soil Sci. Soc.Am. J. 41:966-971.

Harris, Alfred Ray, and Edward A. Hansen. 1975. A new ceramic cup soil-water sampler. Soil Sci. Soc. Amer. Proc. 39:157-158.

Heimovaara, T. J. 1995. Assessing temporal variations in soil water composition with time domain reflectometry. Soil Sci. Soc. Am. J. 59:689-698.

Hoosbeek, M. R., A. Stein, and R. B. Bryant. 1997. Mapping soil degradation. *In* R. Lal, W. H. Blum, C. Valentine, and B. A. Stewart (Eds.), Methods for Assessment of Soil Degradation, CRC Press, New York, pp. 407-422.

Ingvalson, R. D., J. D. Oster, S.L. Rawlins, and G.J. Hoffman. 1970. Measurement of water potential and osmotic potential in soil with a combined thermocouple psychrometer and salinity sensor. Soil Sci. Soc. Amer. Proc. 34:570-74.

Jackson, D. R., F.S. Brinkley, and E.A. Bendetti. 1976. Extraction of soil water using cellulose-acetate hollow fibers. Soil Sci. Soc. Am. J. 40:327-329.

Jaynes, D.B. 1996. Improved soil mapping using electromagnetic induction surveys. *In* P.C. Robert, R.H. Rust, and W.E. Larson (ed.) Proc. 3rd Inter. Conf. on Precision Agriculture. Minneapolis, MN. June 23-26, 1996. ASA, CSSA, SSSA, Madison, WI.

Johnston, M. A. 1994. An evaluation of the four-electrode and electromagnetic induction techniques of soil salinity measurement. Water Research Commission Report No. 269/1/94, South Africa, pp. 191.

Jordan, Carl F. 1968. A simple, tension-free lysimeter. Soil Sci.105:81-86.

Jury, W. A. 1975a. Solute travel-time estimates for tile-drained fields: I. Theory. Soil Sci. Soc. Am. J. 39:1020-1024.

Jury, W. A. 1975b. Solute travel-time estimates for tile-drained fields. II. Application to experimental studies. Soil Sci. Soc. Am. J. 39:1024-1028.

Kaddah, M. T. and J. D. Rhoades. 1976. Salt and water balance in Imperial Valley, California. Soil Sci. Soc. Am. J. 40:93-100.

Kachanoski, R. G., E. G. Gregorich and I. J. van Wesenbeeck. 1988. Estimating spatial variations of soil water content using noncontacting electromagnetic inductive methods. Can. J. Soil Sci. 68: 715-722.

Kemper, W. D. 1959. Estimation of osmotic stress in soil water from the electrical resistance of finely porous ceramic units. Soil Sci. 87:345-349.

Kinniburgh, D.G., and D.L. Miles. 1983. Extraction and chemical analysis of interstitial water from soils and rocks. Environ. Sci. and Technol.17:362-368.

Kitchen, N. R., K. A. Sudduth and S. T. Drummond. 1996. Mapping of sand deposition from 1993 Midwest floods with electromagnetic induction measurements. J. Soil and Water Cons. 51(4):336-340.

Kittrick, J. A. 1983. Accuracy of several immiscible displacement liquids. Soil Sci. Soc.Am. J. 47:1045-1047.

Kohnke, H., F.R. Dreibelbis, and J.M. Davidson. 1940. A survey and discussion of lysimeters and a bibliography on their construction and performance. U.S. Dept. Agr. Misc. Publ. No. 372.

Lesch, S.M., J. D. Rhoades, L.J. Lund and D.L. Corwin. 1992. Mapping soil salinity using calibrated electromagnetic measurements. Soil Sci. Soc. Am. J. 56(2):540-548.

Lesch, S.M., D. J. Strauss and J. D. Rhoades. 1995a. Spatial prediction of soil salinity using electromagnetic induction techniques: 1. Statistical prediction models: A comparison of multiple linear regression and cokriging. Water Resour. Res. 31:373-386.

Lesch, S. M., D. J. Strauss and J. D. Rhoades. 1995b. Spatial prediction of soil salinity using electromagnetic induction techniques: 2. An efficient spatial sampling algorithm suitable for multiple linear regression model identification and estimation. Water Resour. Res. 31: 387-398.

Lesch, S.M., J. D. Rhoades, D. J. Strauss, K. Lin and M.A.A. Co. 1995c. The ESAP user manual and tutorial guide version 1.0. U.S. Salinity Laboratory Research Report No. 138, 108 p.

Lesch, S. M., J. Herrero, and J. D. Rhoades. 1998. Monitoring for temporal changes in soil salinity using electromagnetic induction techniques. Soil Sci. Soc. Amer. J. 62:232-242.

Levin, M.J., and D. R. Jackson. 1977. A comparison of *in situ* extractors for sampling soil water. Soil Sci. Soc. Am. J. 41:535-536.

Litaor, M. Iggy. 1988. Review of soil solution samplers. Water Resour. Res.

24:727-733.

Long, D.S., S. D. DeGloria, and J.M. Galbraith. 1991. Use of the global positioning system in soil survey. J. Soil Water Conserv. 46:293-297.

Longenecker, D. E., and P.J. Lyerly. 1964. Making soil pastes for salinity analysis: A reproducible capillary procedure. Soil Sci. 97:268-275.

Lopez-Bruna, D. and J. Herrero. 1996. The behavior of the electromagnetic sensor and its calibration for soil salinity. Agronomie 16:95-105. (in Spanish)

Loveday, J. 1972. Moisture content of soils for making saturation extracts and effect of grinding. Division of Soils Technical Paper No. 12a, Commonwealth Scientific and Industrial Research Organization, Canberra City, A.C.T. 2601, Australia.

Loveday, J. 1980. Experiences with the 4-electrode resistivity technique for measuring soil salinity. Div. Soils, CSIRO, Rep. 51.

Luke, R. and P. Shaw. 1994a. An economic assessment of the cost of salinity to agriculture in the Campaspe dryland sub-region. Victorian Department of Agriculture, East Melbourne, Australia. 40 pp.

Luke, R. and P. Shaw. 1994b. An economic assessment of the cost of salinity to agriculture in the Loddon dryland sub-region. Victorian Department of Agriculture, East Melbourne, Australia. 40 pp.

Luke, R. and P. Shaw. 1994c. An economic assessment of the cost of salinity to agriculture in the Avoca dryland sub-region. Victorian Department of Agriculture, East Melbourne, Australia. 44 pp.

Maas, E.V. 1986. Salt tolerance of plants. Applied Agricultural Research. 1:12-26.

Maas, E.V. 1990. Chapter 13. Crop salt tolerance. *In* K.K. Tanji (ed.) Agri. salinity assessment and management, ASCE Manuals & Reports on Engineering No. 71. ASCE, N. Y., pp. 262-304.

Maas, E.V., and G.J. Hoffman. 1977. Crop salt tolerance - current assessment. J. Irrig. and Drainage Div., ASCE 103(IR2):115-134.

McKenzie, R.C., W. Chomistek, and N. F. Clark. 1989. Conversion of electromagnetic inductance readings to saturated paste extract values in soils for different temperature, texture, and moisture conditions. Can. J. Soil Sci. 69:25-32.

McKenzie, R.C., D. R. Bennett, and K.M. Riddel. 1990. Mapping salinity to measure potential yield of wheat. Annual Alberta Soil Science Workshop, Edmonton, Canada, pp.268-273.

McNeill, J. D. 1980. Electromagnetic terrain conductivity measurement at low induction numbers. Technical Note TN-6, Geonics Limited, Mississauga, Ontario, Canada.

McNeill, J. D. 1990. Personal communication.

McNeill, J. D. 1992. Rapid, accurate mapping of soil salinity by electromagnetic ground conductivity meters. *In* G.C. Topp, W. D. Reynolds and R. E. Green (eds.) Advances in measurement of soil physical properties: Bringing theory into practice. SSSA Spec. Publ. 30. ASA, CSSA and SSSA, Madison, WI., pp. 209-229.

Menzies, N. W., and L.C. Bell. 1988. Evaluation of the influence of sample preparation and extraction technique on soil solution composition. Aust. J. Soil Res. 26:451-464.

Mubarak, A., and R.A. Olsen. 1976. Immiscible displacement of the soil solution by centrifugation. Soil Sci. Soc. Am. J. 40:329-331.

Mubarak, A., and R.A. Olsen. 1977. A laboratory technique for appraising in situ salinity of soil. Soil Sci. Soc. Am. J. 41:1018-1020.

Nadler, A. 1981. Field application of the four-electrode technique for determining soil solution conductivity. Soil Sci. Soc. Am. J. 45:30-34.

Nadler, A. 1982. Estimating the soil water dependence of the electrical conductivity soil solution/electrical conductivity bulk soil ratio. Soil Sci. Soc. Am. J. 46:722-726.

Nadler, A., and S. Dasberg. 1980. A comparison of different methods for measuring soil salinity. Soil Sci. Soc. Am. J. 44:725-728.

Nadler, A., M. Magaritz, Y. Lapid, and Y. Levy. 1982. A simple system for repeated soil resistance measurements at the same spot. Soil Sci. Soc. Am. J. 46:661-663.

Nielsen, D. R., J.W. Biggar, and K.T. Fah. 1973. Spatial variability of field-measured soil-water properties. Hilgardia. 42:215-259.

Oldeman, L.R., van Engelen, V.W.P., and Pulles, J.H.M. 1991. The extent of human-induced soil degradation. In. Oldeman, L.R., Hakkeling, R.T.A., and Sombroek, W.G. *World Map of the Status of Human-Induced Soil Degradation: An Explanatory Note*. Wageningen: International Soil Reference and Information Centre (ISRIC). 34 pp.

Oliver, M., S. Wilson, J. Gomboso, and T. Muller. 1996. The costs of salinity to government agencies and public utilities in the Murray-Darling basin. ABARE Research Report # 96.2. Australian Bureau of Agricultural and Resource Economics. 83 pp.

Oster, J. D., and R. D. Ingvalson. 1967. *In situ* measurement of soil salinity with a sensor. Soil Sci. Soc. Amer. Proc. 31:572-574.

Oster, J. D., and L. S. Willardson. 1971. Reliability of salinity sensors for the management of soil salinity. Agron. J. 63:695-698.

Oster, J. D. and J. D. Rhoades. 1990. Steady state root zone salt balance. *In* Agricultural Salinity Assessment and Management Manual. K. K. Tanji (ed.). ASCE Manuals & Reports on Engineering No. 71, ASCE, New York. pp. 469-481.

Oster, J. D., L. S. Willardson, and G.J. Hoffman. 1973. Sprinkling and ponding techniques for reclaiming saline soils. Trans. ASCE. 16:89-91.

Oster, J. D., L. S. Willardson, J. van Schilfgaarde, and J. O. Goertzen. 1976. Irrigation control using tensiometers and salinity sensors. Trans. ASAE. 19:294-298.

Parizek, Richard R., and Burke E. Lane. 1970. Soil-water sampling using pan and deep pressure-vacuum lysimeters. J. Hydrol. 11:1-21.

Paul, J. L., K. K. Tanji and W. D. Anderson. 1966. Estimating soil sand saturation extract composition by a computer method. Soil Sci. Soc. Am. Proc. 30:15-17.

Reeve, R.C., and E.J. Doering. 1965. Sampling the soil solution for salinity appraisal. Soil Sci. 99:339-344.

Reitemeier, R.F. 1946. Effect of moisture content on the dissolved and exchangeable ions of soils of arid regions. Soil Sci. 61:195-214.

Rengasamy, P. 1997. Sodic soils. *In R.* Lal, W. H. Blum, C. Valentine, and B. A. Stewart (eds.), Methods for Assessment of Soil Degradation, CRC Press, pp. 265-277.

Rhoades, J. D. 1972. Quality of water for irrigation. Soil Sci. 113:277-284.

Rhoades, J. D. 1976. Measuring, mapping and monitoring field salinity and water table depths with soil resistance measurements. FAO Soils Bulletin. 31:159-186.

Rhoades, J. D. 1978. Monitoring soil salinity: A review of methods. Establishment of water quality monitoring programs. L.G. Everett, K.D. Schmidt (eds.) Am. Water Resources Assoc., San Francisco, CA, June 1978. 2:150-165.

Rhoades, J. D. 1979a. Salinity management and monitoring. Proc. 12th Biennial Conference on Ground Water, Sacramento, CA, Sept. 1979. Calif. Water Resources Center. Report. 45:73-87.

Rhoades, J. D. 1979b. Inexpensive four-electrode probe for monitoring soil salinity. Soil Sci. Soc. Am. J. 43:817-818.

Rhoades, J. D. 1980. Determining leaching fraction from field measurements of soil electrical conductivity. Agric. Water Mgt. 3:205-215.

Rhoades, J. D. 1981. Predicting bulk soil electrical conductivity vs. saturation paste extract electrical conductivity calibrations from soil properties. Soil Sci. Soc. Am. J. 45:43-44.

Rhoades, J. D. 1982. Soluble Salts. *In* A. L. Page, R.H. Miller and D. R. Kenney (eds.) Methods of soil analysis part 2, chemical and microbiological properties. Agronomy Monograph. 9:167-178.

Rhoades, J. D. 1984. Principles and methods of monitoring soil salinity. *In* Soil salinity and irrigation - processes and management. Springer Verlag, Berlin. 5:130-142.

Rhoades, J. D. 1990a. Sensing soil salinity problems: New technology. *In* R.L. Elliott (ed.) Proc. 3rd. National irrigation symposium irrigation association and ASCE, October 28 - November 1, 1990. Phoenix, AZ, pp. 422-428.

Rhoades, J. D. 1990b. Determining soil salinity from measurements of electrical conductivity. Commun. Soil Sci. Plant Anal. 21:861-901.

Rhoades, J. D. 1992a. Instrumental field methods of salinity appraisal. *In* G.C. Topp, W. D. Reynolds and R. E. Green (eds.), Advances in measurement of soil physical properties: Bringing theory into practice. SSSA Spec. Publ. 30. ASA, CSSA and SSSA, Madison, WI., pp. 231-248.

Rhoades, J. D. 1992b. Recent advances in the methodology for measuring and mapping soil salinity. Proc. Int'l Symp. on Strategies for Utilizing Salt Affected Lands, ISSS Meeting, Bangkok, Thailand, Feb. 17-25, 1992.

Rhoades, J. D. 1993. Electrical conductivity methods for measuring and mapping soil salinity. *In*: D. L. Sparks (ed.), Advances in Agonomy. Vol. 49:201-251.

Rhoades, J. D. 1994. Soil salinity assessment: Recent advances and findings. ISSS Sub-Commission A Meeting, Acupulco, Mexico, July 10-16, 1994.

Rhoades, J. D. 1996a. Salinity: Electrical conductivity and total dissolved salts. In Methods of Soil Analysis, Part 3, Chemical Methods, Soil Sci. Soc. Am. Book Series 5, Amer. Soc. of Agron., Inc., Madison, Wisconsin, pp. 417-435.

Rhoades, J. D. 1996b. New Assessment technology for the diagnosis and control of salinity in irrigated lands. Proc. International Symposium on Development of Basic Technology for Sustainable Agriculture Under Saline Conditions, Arid Land Research Center, Tottori University, Tottori, Japan, December 12, 1996. pp. 1-9.

Rhoades, J. D. 1997a. Sustainability of Irrigation: An Overview of Salinity Problems and Control Strategies. Proc. 1997 Annual Conference of the Canadian Water Resources Association, Footprints of Humanity: Reflections on Fifty Years of Water Resources Developments, Lethbridge, Alberta, Canada, June 3-6, 1997., pp. 1-40.

Rhoades, J. D. 1997b. Geospatial measurements of soil electrical conductivity to determine soil salinity and diffuse salt-loading from irrigation. Proc. Joint AGU Chapman Conference/SSSA Outreach Conference on the Application of GIS, Remote Sensing, Geostatistics, and Solute Transport Modeling to the Assessment of Nonpoint Source Pollutants in the Vadose Zone. (in press).

Rhoades, J. D. 1998. Use of Saline and Brackish Waters for Irrigation: Implications and Role in Increasing Food Production, Conserving Water, Sustaining Irrigation and Controlling Soil and Water Degradation. *In* R. Ragab and G. Pearce (eds.) Proceedings of the International Workshop on "The use of saline and brackish water for irrigation - implications for the management of irrigation, drainage and crops", Bali, Indonesia, pages 23-24.

Rhoades, J. D., and R .D. Ingvalson. 1971. Determining salinity in field soils with soil resistance measurements. Soil Sci. Soc. Amer. Proc. 35:54-60.

Rhoades, J. D. and S. D. Merrill. 1976. Assessing the suitability of water for irrigation: Theoretical and empirical approaches. FAO Soils Bulletin 31:69-109.

Rhoades, J. D., and J. van Schilfgaarde. 1976. An electrical conductivity probe for determining soil salinity. Soil Sci. Soc. Am. J. 40:647-651.

Rhoades, J. D., and D.L. Corwin. 1981. Determining soil electrical conductivity - depth relations using an inductive electromagnetic soil conductivity meter. Soil Sci. Soc. Am. J. 45:255-260.

Rhoades, J. D., and D.L. Corwin. 1984. Monitoring soil salinity. J. Soil and Water Conservation 39 (3) : 172-175.

Rhoades, J. D., and J. D. Oster. 1986. Solute content. *In* A. Klute (ed.) Methods of soil analysis, Part II. Physical and Mineralogical Methods 2nd Ed. Am. Soc. Agron. Monogr. 9:985-1006, Madison, WI.

Rhoades, J. D., and D.L. Corwin. 1990. Soil electrical conductivity: Effects of soil properties and application to soil salinity appraisal. Commun. Soil Sci. Plant Anal. 21:837-860.

Rhoades, J. D., and S. Miyamoto. 1990. Testing soils for salinity and sodicity. Soil Testing and Plant Anal. 3rd Ed., SSSA Book Series No. 3, pp. 299-336.

Rhoades, J. D., J. D. Oster, R. D. Ingvalson, J. M. Tucker and M. Clark. 1974. Minimizing the salt burdens of irrigation drainage waters. J. Environ. Qual. 3:311-316.

Rhoades, J. D., P. A. C. Raats, and R.J. Prather. 1976. Effects of liquid-phase electrical conductivity, water content, and surface conductivity on bulk soil electrical conductivity. Soil Sci. Soc. Am. J. 40:651-655.

Rhoades, J. D., M. T. Kaddah, A. D. Halvorson, and R.J. Prather. 1977. Establishing soil electrical conductivity-salinity calibrations using four-electrode cells containing undisturbed soil cores. Soil Sci. 123:137-141.

Rhoades, J. D., D.L. Corwin, and G.J. Hoffman. 1981. Scheduling and controlling irrigations from measurements of soil electrical conductivity. Proc. ASAE Irrigation Scheduling Conference, Chicago, IL., Dec. 14, 1981. pp. 106-115.

Rhoades, J. D., N. A. Manteghi, P.J. Shouse, and W.J. Alves. 1989a. Soil electrical conductivity and soil salinity: New formulations and calibrations. Soil Sci. Soc.Am. J. 53:433-439.

Rhoades, J. D., N. A. Manteghi, P.J. Shouse, and W.J. Alves. 1989b. Estimating soil salinity from saturated soil-paste electrical conductivity. Soil Sci. Soc. Am. J. 53:428-433.

Rhoades, J. D., B.L. Waggoner, P.J. Shouse, and W.J. Alves. 1989c. Determining soil salinity from soil and soil-paste electrical conductivities: Sensitivity analysis of models. Soil Sci. Soc. Am. J. 53:1368-1374.

Rhoades, J. D., S.M. Lesch, P.J. Shouse, and W.J. Alves. 1989d. New calibrations for determining soil electrical conductivity - depth relations from electromagnetic measurements. Soil Sci. Soc. Am. J. 53:74-79.

Rhoades, J. D., P.J. Shouse, W.J. Alves, N. A. Manteghi, and S.M. Lesch. 1990a. Determining soil salinity from soil electrical conductivity using different models and estimates. Soil Sci. Soc. Am. J. 54:46-54.

Rhoades, J. D., D.L. Corwin, and S.M. Lesch. 1990b. Effect of soil EC_a - depth profile pattern on electromagnetic induction measurements. U.S. Salinity Lab. Report. #125, 108 pp.

Rhoades, J. D., S. M. Lesch, P. J. Shouse and W. J. Alves. 1990c. Locating sampling sites for salinity mapping. Soil Sci. Soc. Am. J. 54:1799-1803.

Rhoades, J. D., A. Kandiah and A. M. Mashali. 1992. The use of saline waters for crop production. FAO Irrigation & Drainage Paper 48, FAO, Rome, Italy. 133 p.

Rhoades, J. D., S. M. Lesch, R. D. LeMert and W. J. Alves. 1997a. Assessing irrigation/drainage/salinity management using spatially referenced salinity measurements. Agr. Water Mgt. 35:147-165.

Rhoades, J. D., S. M. Lesch, S. L. Burch, R. D. LeMert, P. J. Shouse, J. D. Oster and T. O'Halloran. 1997b. Salt transport in cracking soils: Salt distributions in soil profiles and fields and salt pick-up by run-off waters. J. Irrig. & Drainage Engineering, ASCE 123:323-328.

Rhoades, J. D., N. A. Manteghi, S. M. Lesch and D. C. Slovacek. 1997c. Determining soil & water sodicity from electrode measurements. Commun. Soil Sci. Plant Anal. 28:1737-1765.

Rhoades, J. D., S. M. Lesch and D. B. Jaynes. 1997d. Geospatial measurements of soil electrical conductivity for prescription farming. Special Publ. Soil Sci. Soc. Am., ASA, Madison, WI. (in press).

Richards, L. A. 1941. A pressure-membrane extraction apparatus for soil solution. Soil Sci. Soc. 51:377-386.

Richards, L. A. 1966. A soil salinity sensor of improved design. Soil Sci. Soc. Amer. Proc. 30:333-337.

Rose, C. W., P. W. A. Dayananda, D. R. Nielsen and J. M. Biggar. 1979. Long-term solute dynamics and hydrology in irrigated slowly permeable soils. Irrig. Sci. 1:77-87.

Ross, D.S., and R.J. Bartlett. 1990. Effects of extraction methods and sample storage on properties of solutions obtained from forested spodosols. J. Environ. Qual. 19:108-113.

Seckler, David. 1996. The New Era of Water Resources Management: From "Dry" to "Wet" Water Savings, Consultative Group on International Agricultural Research, Washington, D C.

Shainberg, I., J. D. Rhoades, and R. J. Prather. 1980. Effect of exchangeable sodium percentage, cation exchange capacity, and soil solution composition on soil electrical conductivity. Soil Sci. Soc. Am. J. 44:469-473.

Shaw, R. 1994. Estimation of the electrical conductivity of saturation extracts from the electrical conductivity of 1:5 soil:water suspensions and various soil properties. Project Report Series QO94025, Dept. of Primary Industries, Queensland, Australia, 41 pp.

Shimshi, Daniel. 1966. Use of ceramic points for the sampling of soil solution. Soil Sci. 101:98-103.

Slavich, P.G. and B. J. Read. 1983. Assessment of electromagnetic induction measurements using an inductive electromagnetic soil conductivity meter. Soil Sci. Soc. Amer. J. 45:255-260.

Slavich, P.G. 1990. Determining EC_a - depth profiles from electromagnetic induction measurements. Aust. J. Soil Research 28:443-452.

Slavich, P.G. and J. Yang. 1990. Estimation of field leaching rates from chloride mass balance and electromagnetic induction measurements. Irrig. Sci. 11:7-14.

Slavich, P. G. and G.H. Peterson. 1990. Estimating average rootzone salinity from electromagnetic induction (EM-38) measurements. Aust. J. Soil Res. 28:453-463.

Soil Science Society of America. 1996. Methods of Soil Analysis, Part 3, Chemical Methods. American Society of Agronomy, Madison, Wisconsin, USA. pp. 1389.

Sonnevelt, C., and J. van den Ende. 1971. Soil analysis by means of a 1:2 volume extract. Plant Soil. 35:505-516.

Suarez, D.L. 1986. A soil water extractor that minimizes CO_2 degassing and pH errors. Water Resour. Res. 22:876-880.

Suarez, D.L. 1987. Prediction of pH errors in soil water extractors due to degassing. Soil Sci. Soc. Am. J. 51:64-67.

SURFER. 1986. Reference manual. Golden Software, Inc., Golden, CO.

Tadros, V. T., and J.W. McGarity. 1976. A method for collecting soil percolate and soil solution in the field. Plant and Soil. 44:655-667.

US Salinity Laboratory Staff. 1954. Diagnosis and improvement of saline and alkali soils. USDA Handbook 60, U.S. Government Printing Office, Washington, D. C.

van De Pol, R.M., P.J. Wierenga, and D. R. Nielsen. 1977. Solute movement in a field. Soil Sci. Soc. Am. J. 41:10-13.

van Hoorn, J.W. 1980. The calibration of four-electrode soil conductivity measurements for determining soil salinity. Proc. Int. Symp. Salt Affected Soils, Karnal, pp. 148-156.

Wagner, George H. 1965. Changes in nitrate N in field plot profiles as measured by the porous cup technique. Soil Sci. 100:397-402.

Webster, R. 1985. Quantitative spatial analysis of soil in the field. Advances in Soil Science. 31:505-524.

Webster, R. 1989. Recent achievements in geostatistical analysis of soil. Agrokemia Es Talajtan. 38:519-536.

Wenner, F. 1916. A method of measuring earth resistivity. U. S. Dept. Com. Cur. Of Stand. Sci. Paper No. 258.

Wesseling, J., and J. D. Oster. 1973. Response of salinity sensors to rapidly changing salinity. Soil Sci. Soc. Amer. Proc. 37:553-557.

Wilcox, L.V. 1951. A method for calculating the saturation percentage from the weight of a known volume of saturated soil paste. Soil Sci. 72:233-237.

Williams, B. G. and G. C. Baker. 1982. An electromagnetic induction technique for reconnaissance surveys of soil salinity hazards. Aust. J. Soil Research 20:107-118.

Wollenhaupt, N. C., J.L. Richardson, J.E. Foss, and E. C. Doll. 1986. A rapid method for estimating weighted soil salinity from apparent soil electrical conductivity with an above ground electromagnetic induction meter. Can. J. Soil Sci. 66:315-321.

Wolt, J., and J.G. Graveel. 1986. A rapid method for obtaining soil solution using vacuum displacement. Soil Sci. Soc. Am. J. 50:602-605.

Wood, J. D. 1978. Calibration stability and response time for salinity sensors. Soil Sci. Soc. Am. J. 42:248-250.

Wood, Warren W. 1973. A technique using porous cups for water sampling at any depth in the unsaturated zone. Water Resour. Res. 9:486-488.

Yadav, B. R., N. H. Rao, K.V. Paliwal, and P. B. S. Sarma. 1979. Comparison of different methods for measuring soil salinity under field conditions. Soil Sci. 127:335-339.

Yamasaki, S., and A. Kishita. 1972. Studies on soil solution with reference to nutrient availability. I. Effect of various potassium fertilizer on its behavior in the soil solution. Soil. Sci. and Plant Nutr. 18:1-6.

Annex 1

Methods for establishing $EC_e = F(Ec_a)$ calibrations

EC_e-EC_a calibrations may be directly established in one of four ways depending on equipment availability, time availability, and desired accuracy.

1. The earliest-used, simplest and least accurate method is to measure EC_a (using a surface array of four-electrodes or EM-38 sensor) at numerous locations in the field (or area) of interest, to analyze for salinity (EC_e) soil samples collected at no less than 4-6 sites that provide an approximately equally spaced interval of EC_a readings over the observed range and to determine the best-fit linear relation between the EC_e and EC_a data-pairs using a graphical plot or standard regression analysis procedures. Examples of calibrations established in this manner are shown in Figures 48 and 49. Since soil salinity is typically quite variable from spot-to spot and with depth in saline fields/areas, numerous samples should be taken within the dimensions of the measurement-volume given in Chapter 3 for the four-electrode and EM-38 sensors, respectively, in order to obtain a representative sample of the relatively large volumes of soil measured by these sensors. For any site, the samples should be composited for each depth-increment of interest. This type of calibration is limited to whatever range of salinity exists in the field (area) at the time of sampling. The accuracy is limited by the variability in soil properties that exist within the sampled area and by the degree to which the relatively small soil sample represents the much larger volume of soil measured by the sensors. The accuracy is generally sufficient for salinity diagnosis and gross mapping purposes, but not for certain other assessment purposes.

2. A more accurate method is depicted in Figure 17. In this method a soil EC-probe is used to determine the EC_a values of small volumes of soil that have been adjusted in the field to provide a desired optimum range and distribution of salinity values. To adjust the salinity, saline waters of various salinities (EC = 5, 10, 20, 40, etc.; SAR values = 8, or whatever else is deemed appropriate) are impounded in column sections (30 cm in diameter by 45 cm in length) driven about 10-15 cm into the soil and in a surrounding 15-cm wide excavated moat (see Figure 17a). (About 40 litres of saline water is required to bring the soil to a depth of 30 cm beneath the impounded area to the desired level of salinity).

   When the soil has drained to about "field-capacity" water content, 2-3 days after the impounded water has infiltrated, a 2.5 cm hole is cored to a depth of 30 cm in the centre of the uniformly salinized soil volume using a Lord, or equivalent diameter coring tube (see Figure 17b). Next, the soil EC-probe is inserted (see Figure 17c) into the slightly undersized hole to a depth that centres its electrodes at the desired depth (usually 15 cm) and the reading of EC_a is obtained along with the temperature (using either the thermistor in the Martek probe or any suitable temperature probe). After the soil EC-probe is removed, the soil immediately exterior to the position of the centered EC-probe is sampled (essentially the 7.5-22.5 cm depth) using a 10-15 cm diameter soil auger (see Figure 17d). This sample of soil (which closely correspond to the volume measured by the EC-probe for EC_a) is analysed by any conventional method for salinity (EC_e). The linear EC_e-EC_a calibration is established by graphical means or by regression analysis. Example calibrations of this type are shown in Figures 19 and 20. Such calibrations are the quickest to obtain of any of the methods and are generally quite accurate because the volumes of soil sampled for EC_e and EC_a are nearly the same and because the variability in soil type and water content is minimized, as is salinity (within the sampled region of soil). Such calibrations have been used to establish predictive calibration-relations by soil type using auxiliary soil-property data of the soil samples obtained during the calibrations of numerous different soils. Very nice relations have been established in this regard as previously discussed in the main text sections of this report.

FIGURE 1.1
Series of four-electrode cells containing undisturbed soil-cores being segmented after removal from the soil-core sampler

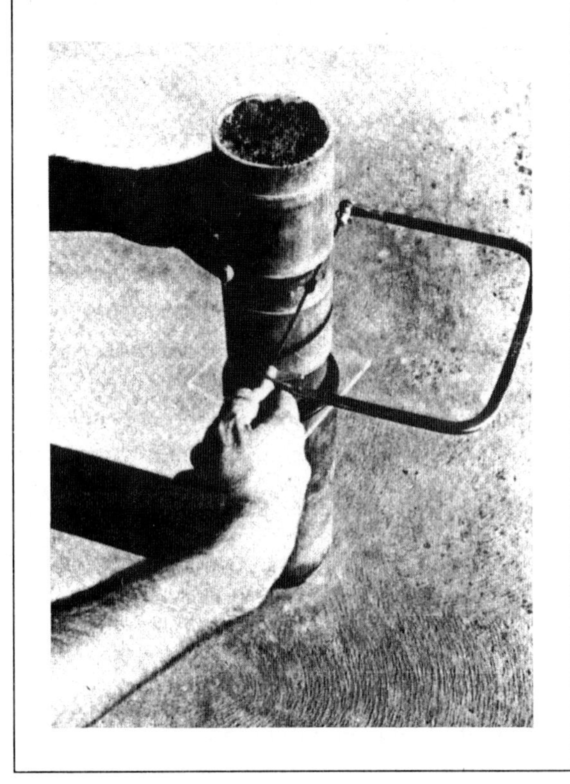

3. A still more accurate direct calibration method has been developed using specially built four-electrode cells that fit as removable inserts within soil coring devices used to obtain undisturbed soil-core samples. Undisturbed soil cores are collected from the soil sites representative of the soil type, field, or area of interest (or from the soil bodies whose salinity has been adjusted by the previously described method) using lucite column sections as corer-inserts.. A soil-filled four-electrode cell is obtained by slicing through the soil core, removed from the corer, between adjacent cylinder segments and then screwing electrodes into the tapped hoes in the cell wall, as illustrated in Figures 18 and 1.1. The composite EC_a of the soil is calculated from the succession of readings made by sequentially connecting the four-electrode generator/meter to the eight combinations of electrodes that can be achieved with the eight equidistant-spaced electrodes inserted into the cell/soil (only four electrodes are shown in Figure 18). The cell constant (k) of these four-electrode cells are obtained by filling them with standard EC-solutions and measuring the resistance of the filled-cell using Equations [1] to [3]. The soil temperature is also measured using a suitable temperature probe and EC_a is calculated from the these data using Equations [1] to [3]. After EC_a is measured, the soil is removed from the cell and analyzed for salinity (and other properties of interest). Next, the linear EC_e-EC_a calibration is established as described for the other methods. These types of calibrations are very

accurate because the measurements of EC_e and EC_a are made on exactly the same volume of soil. An example of the improvement in accuracy is illustrated in Figure 1-2; the improvement in the lower range of salinity is especially evident. The above-described methods of calibration are described in more detail elsewhere (Rhoades, 1976; Rhoades and Halvorson, 1977; Rhoades *et al.*, 1977).

4. An analogous but more accurate and statistically rigorous method for establishing EC_e-EC_a calibrations that apply to specific field situations is the stochastic, field-calibration method previously described in the main text.

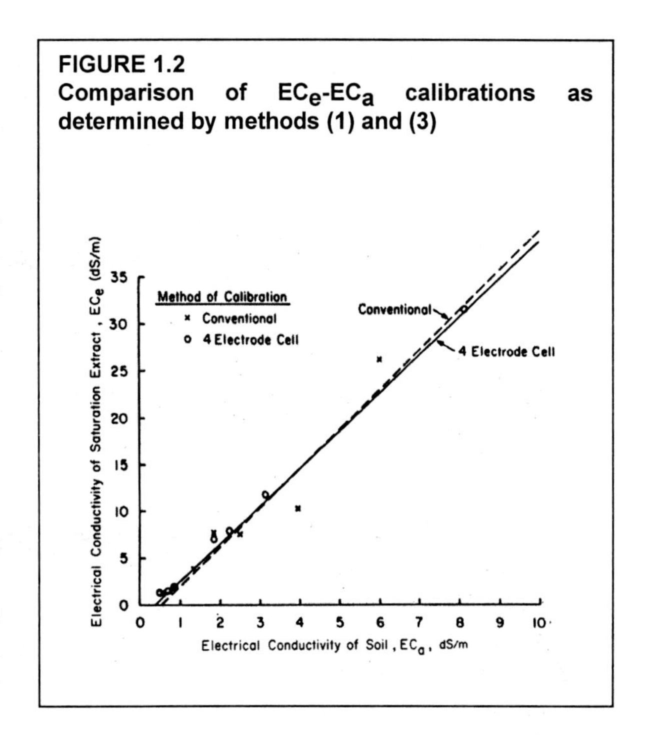

FIGURE 1.2
Comparison of EC_e-EC_a calibrations as determined by methods (1) and (3)

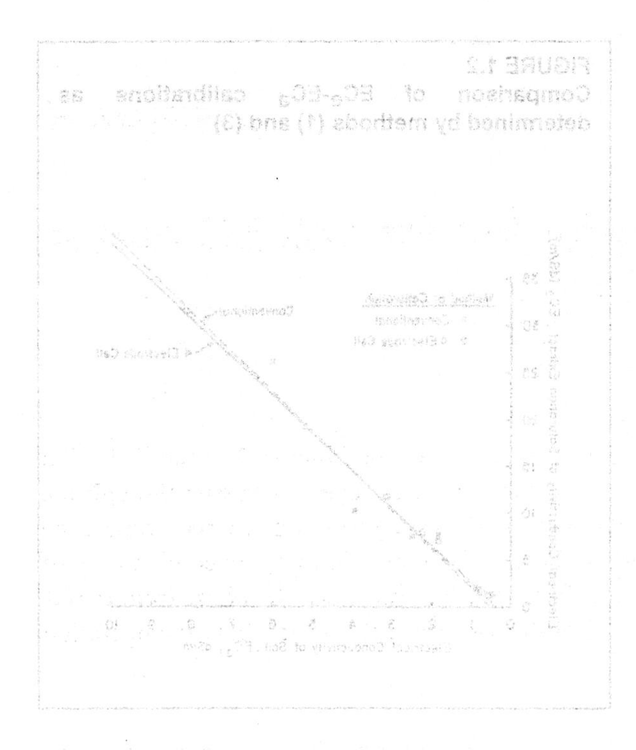

Annex 2

Circuitry and parts-list for soil EC-meter

The circuit developed for use with soil EC-probes is shown in Figures 2.1 and 2.2. The components cost is less than about US$100. With this circuit, a current is passed through the outside pair of current electrodes, and the voltage across the inside pair is measured. Since, the current and voltage are known, the resistance can be calculated and, in turn, the conductivity of the soil determined using Equation [24]. The accuracy of the circuit is ± 0.3 ohms from 0-100 ohms and ± 4 ohms from 100-200 ohms (or 2 % of range). For more detail about the operation of the circuit see Austin and Rhoades (1979).

FIGURE 2.1
Diagram of a simplified circuit or EC_a-meter; R_y, the resistance of the null adjustment potentiometer, is adjusted to equal R_x, the resistance of the salinity sensor (after Austin and Rhoades, 1979)

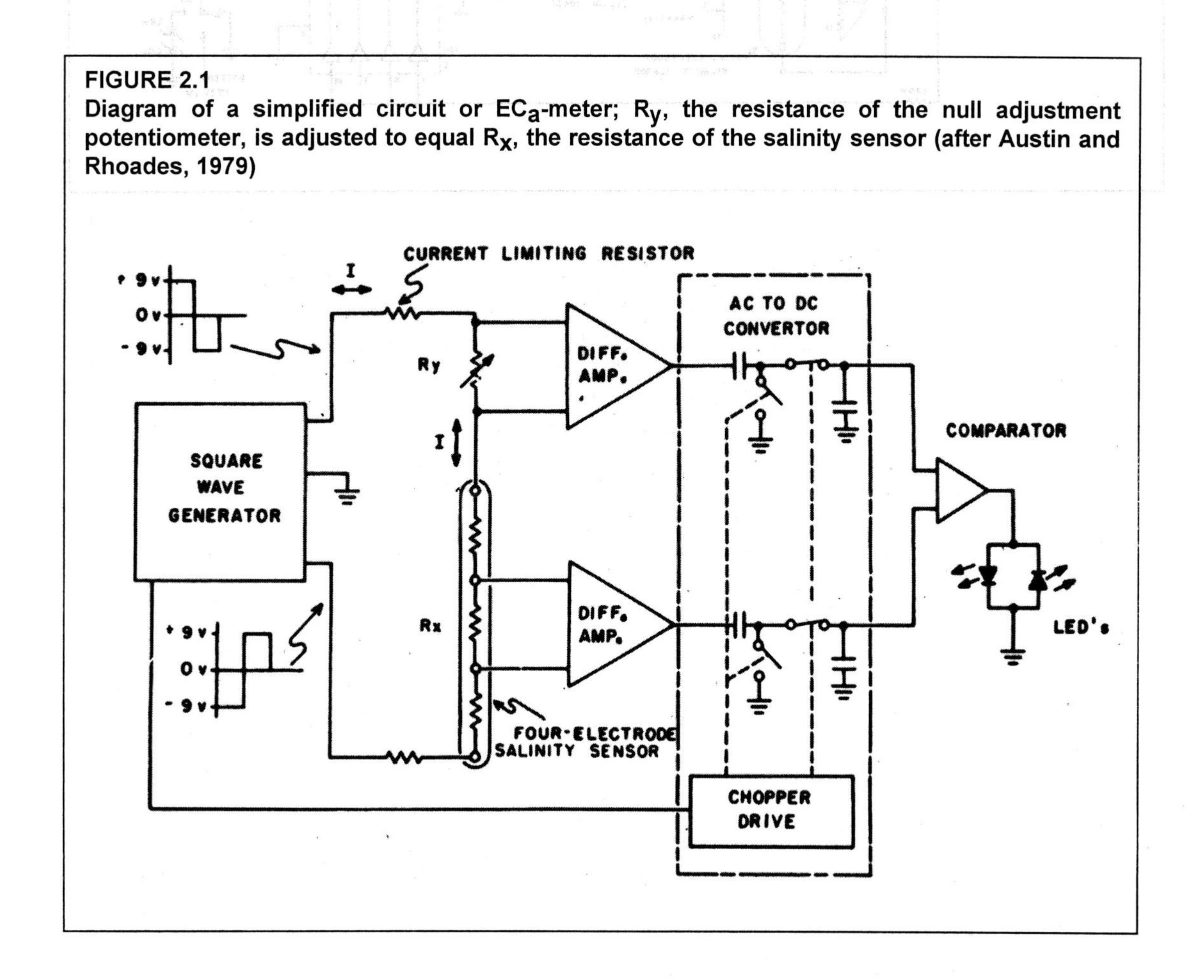

FIGURE 2.2
Detailed diagram of low-cost circuit for reading four-electrode sensors (after Austin and Rhoades, 1979)

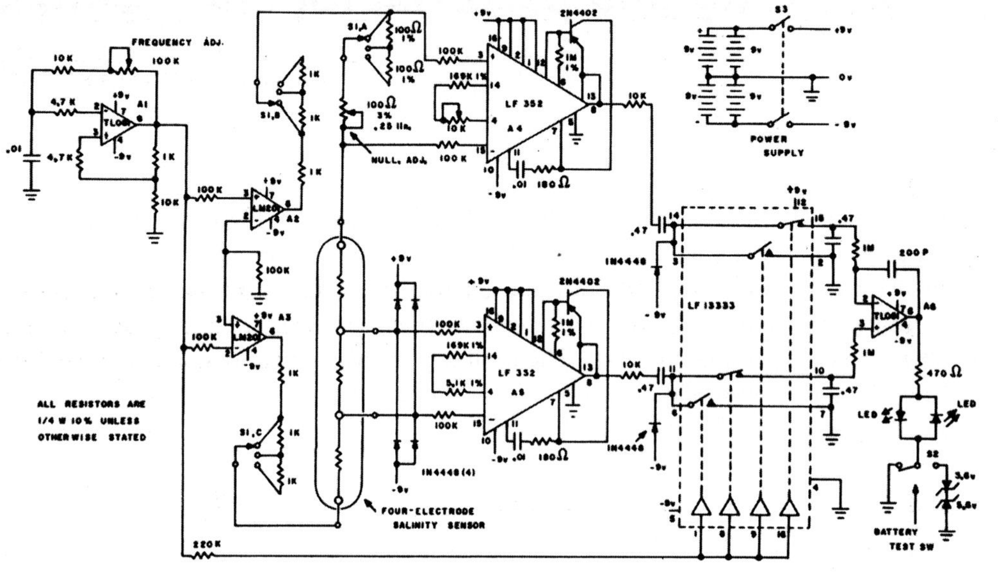

ALL RESISTORS ARE 1/4 W 10% UNLESS OTHERWISE STATED

Annex 3

Equation for calculating effect of insertion-depth of four-electrodes

The depth of electrode-insertion affects the measurement of soil electrical conductivity. For electrodes configured equidistantly apart in the so-called Wenner-array, the relationship between EC_a, distance between electrodes (a, in m) and depth of electrode insertion (b, in m) is, according to Wenner (1916), as follows:

$$EC_a = \left(1 + \frac{2}{\sqrt{1+4(b/a)^2}} - \frac{1}{\sqrt{1+(b/a)^2}}\right)/(4\pi a R), \qquad [43]$$

where R is resistance in ohms. The term inside the brackets in the numerator approaches the limiting value of 2 as the depth of electrode insertion (b) becomes small relative to the distance between the electrodes (a).

Annex 4

Construction of burial-type four-electrode probe

Salinity monitoring sometimes requires that repeated measurements be made over a period of time at the same location. For such uses, implanted probes offer certain advantages. For this reason, an inexpensive four-electrode unit was developed that can be implanted and left in the soil for extended periods of time. This burial-type probe is constructed using the components shown in Figure 4.1.

FIGURE 4.1
Inexpensive four-electrode burial-type probe before and after assembly (after Rhoades, 1979)

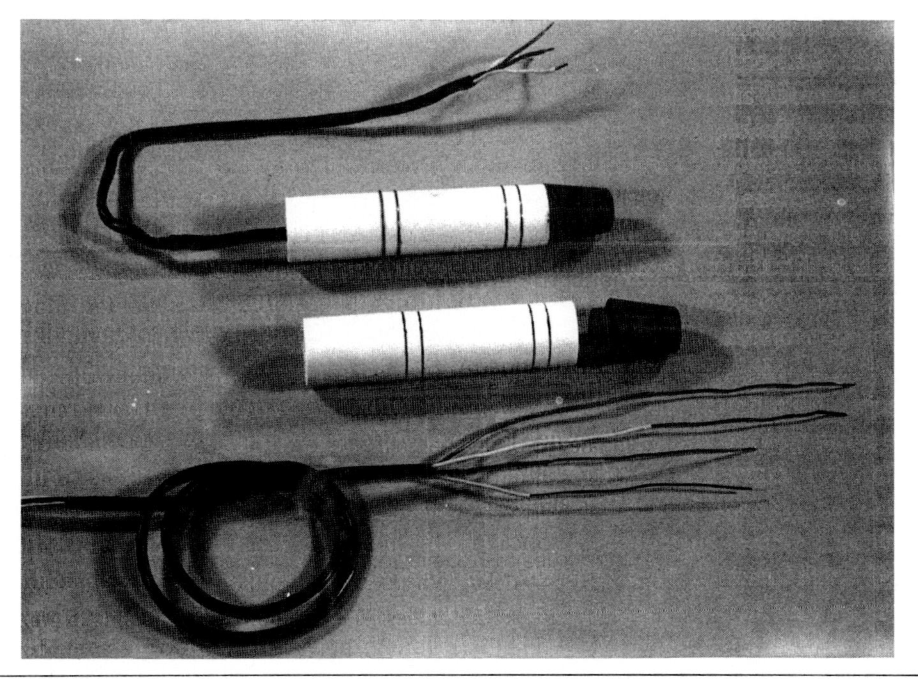

The probe casing is a 4 3/4 inch long length of PVC pipe (3/4 inch schedule 80) in which four grooves (0.040 inch wide and 0.025 inch deep) are made. The distance between the outside grooves is 2.5 inches; the inside grooves are inset by 0.25 inches. The outside wall of the pipe is tapered at 0.5° on a lathe. Two holes (1/16 inch in diameter) are drilled in each of the grooves for the passage of 18-gauge thermostat-type wire, with the insulation removed from the last 4 inches. Each of the four bare wire-ends is passed from inside of the pipe through a hole, is rapped around the outside of the pipe in one of the grooves to form one of the four electrodes, and is returned to the inside of the pipe through a second hole in each groove. The returned end of the wire is crimped against the inside pipe wall to firmly secure the wire in place. A 1.37 inch-long tapered (7° on a side) solid PVC tip is inserted into the head-end of the probe and cemented in-place with PVC solvent cement to serve as a leading edge. Finally, the inside of the pipe is filled with laminating resin to within 0.75 inches of the exit end in order to prevent water entry. The materials cost is less than about US$1. per probe. Data illustrating the utility of the burial-probe are given in Figures 80, 81, and 82. More details about the construction of these probes and about devices to facilitate their installation in the soil are given in Rhoades (1979). The commercialized version of the probe, which consists in part of plastic and which is formed by a pressure injection/molding techniques, is shown in Figure 38.

Annex 5

Examples of various special-use four-electrode cells and sensors

Various types of four-electrode cells and miniaturized four-electrode sensors have been developed for special needs. A few are shown here to illustrate how easy they are to make and to provide ideas for their construction, as well as for others. The detailed information is not provided, since the exact dimensions vary with purpose and availability of materials. Figures 5.1 and 5.2 illustrate a four-electrode cell that permits the composition and water content in soil cores (either packed or undisturbed cores) to be varied and the corresponding value of soil electrical conductivity to be determined (after Bottraud and Rhoades, 1985b). Two other versions of cells used to adjust soil columns to different solution compositions, with and without the opportunity for changes in soil volume, are shown in Figure 5.3 (after Bottraud and Rhoades, 1985a). Analogous versions shown in Figure 5.4 have been made with ceramic plates as bases to permit the application of pressure so as to be able to vary water content over a greater range than possible with the cells shown in Figures 5.1 and 5.2. Even greater pressures can be applied to vary water content using pressure plate apparatus and the types of cells and undisturbed soil cores. Micro four-electrode sensors, such as those shown in Figures 5.5 and 5.6, can be constructed to permit the measurement on EC_a in very shallow soil depths (Figure 5.6) or in small increments of an exposed (excavated) soil profile (Figure 5.5).

FIGURE 5.1
Schematic of apparatus used to vary water content and composition and of a four-electrode cell used to measure the associated values of soil electrical conductivity (after Bottraud and Rhoades, 1985b)

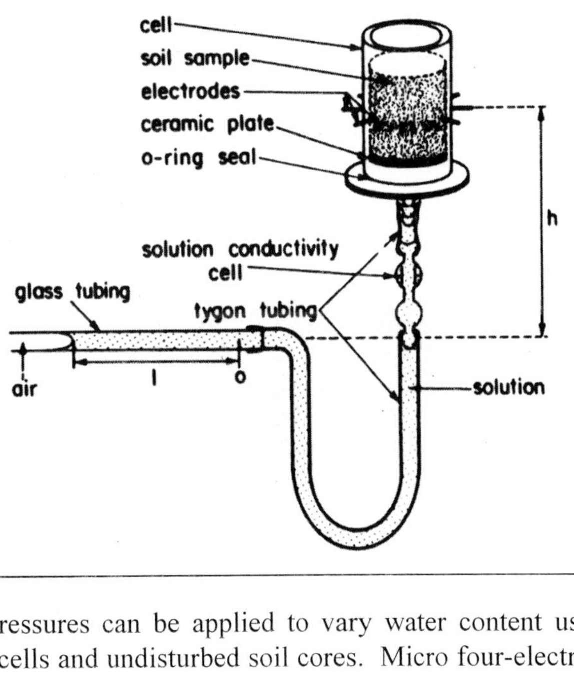

FIGURE 5.2
Several kinds of four-electrode cells and bases used to vary water content and composition and to measure the associated values of soil electrical conductivity

FIGURE 5.3
Schematic of apparatus used to hold four-electrode cells for studying the effects of varying solution composition and induced changes in water content and porosity and to measure the associated values of soil electrical conductivity (after Bottraud and Rhoades, 1985a)

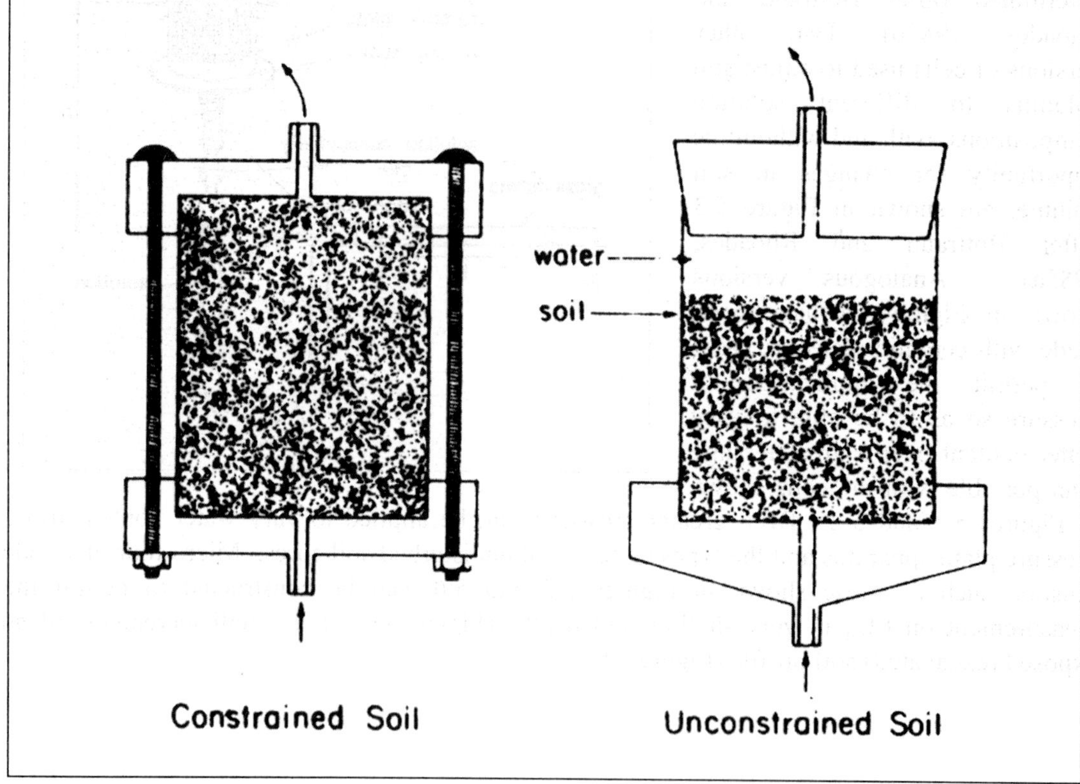

FIGURE 5.4
An apparatus built to hold a four-electrode cell with ceramic containing end-plates to permit changes in water content to be induced by the application of pressure

FIGURE 5.5
Small-span four-electrode sensor used to measure soil electrical conductivity in small depth-increments along an exposed soil profile

FIGURE 5.6
(A) Small-span four-electrode array used to measure soil electrical conductivity in shallow depths, and (B) connection of meter to four-electrode cell (after Rhoades *et al.*, 1977)

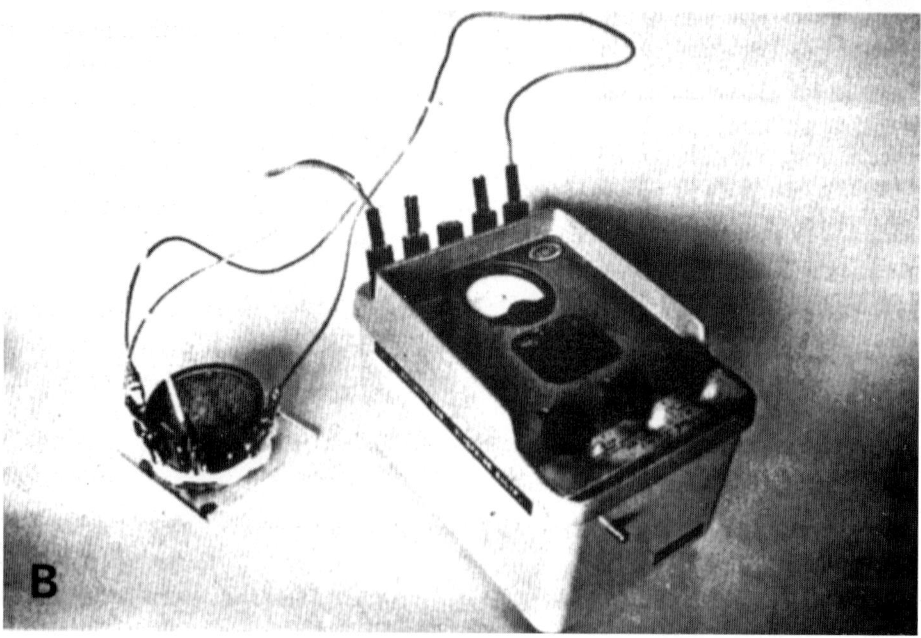

Annex 6

Derivation of EM_{adj} equations

This appendix gives the original basis of the relations of the type given in Equation [22] and Table1 for estimating the EC_a levels for different depth-intervals of the soil profile using only two EM-38 readings. In this approach the EM-38 readings are related to EC_a by a series of simple equations which are based, in part, upon the theoretical response functions of the sensor for homogeneous media as given in Figure 44. This figure shows that EM_H and EM_V measurements give depth-weighted EC_a values to about 1 and 2 metres, respectively. For the 0 to 0.3-m increment of soil, the following relations apply for homogeneous profiles:

$$EM_{0,V} = 0.150\ EC_{0\text{-}0.3,\ V} + 0.850\ EC_{>0.3,V} \text{ and} \qquad [a]$$

$$EM_{0,H} = 0.435\ EC_{0\text{-}0.3,\ H} + 0.565\ EC_{>0.3,\ H} \qquad [b]$$

where $EM_{0,V}$ and $EM_{0,H}$ are the EM-38 values measured at the soil surface in the vertical and horizontal positions, respectively; and $EC_{0\text{-}0.3,\ V}$, $EC_{>0.3,V}$, $EC_{0\text{-}0.3,\ H}$, and $EC_{>0.3,\ H}$ are the actual EC_a values for the 0 to 0.3-m and >0.3-m soil depth intervals. He subscript "$_a$ " in EC_a is dropped from these equations and some of the following ones in order to minimize the clutter in the subscripts. In an homogeneous profile, the 0 to 0.3-m depth of soil only contributes 15 % of the $EM_{0,V}$ value, while the deeper depths contribute 85 %. The corresponding values for the $EM_{0,H}$ value are 43.5 % and 56.5 %, respectively.

Since the volume of soil measured within the 0 to 0.3-m increment is very similar for the vertical and horizontal orientations, it is reasonable to assume that $EC_{0\text{-}0.3,\ V} = EC_{0\text{-}0.3,\ H}$. However, in the case of the >0.3-m increment, the volumes of measurement are quite different and, consequently, $EC_{>0.3,V}$ and $EC_{>0.3,\ H}$ can not be assumed to be equal in value. However, in order to establish a relationship between $EC_{a,\ 0\text{-}0.3}$, $EM_{0,V}$ and $EM_{0,H}$ using equations [a] and [b], it is necessary to equate $EC_{>0.3,V}$ and $EC_{>0.3,\ H}$. This problem was overcome when it was found using empirical data that $EM_{0,H}$ could be adjusted using empirical relationships (see Figure 6.1) so that $EC_{>0.3,V}$ calculated from Equation [a] would equal $EC_{>0.3,\ H}$ calculated from Equation [b]. The empirical data and its means of collection are explained in detail in Corwin and Rhoades (1982, 1990). Briefly, an adjusted $EM_{0,H}$ (for the 0 to 0.3-m increment) was calculated from Equation [b] using the measured values of $EC_{a,\ 0\text{-}0.3}$ and the values of $EC_{>0.3,V}$ calculated from Equation [a]. The plot of measured and adjusted $EM_{0,H}$ values for each depth increment of the test soils revealed the set of linear relations shown in Figure 6.1. Assuming these relations would apply to other soils, measured values of $EM_{0,H}$ for a specified depth increment (0-h metres) can be adjusted so that $EC_{>h,\ V} = EC_{>h,\ H}$, as was demonstrated for the 0 to 0.3-m depth-increment as follows:

FIGURE 6.1
Relationship between electromagnetic soil conductivity as measured by the EM-38 in the horizontal position at the soil surface, $EM_{0,\,H\,(measured)}$ and adjusted electromagnetic soil conductivity, $EM_{0,\,H\,(adjusted)}$ for composite depths (after Corwin and Rhoades, 1990)

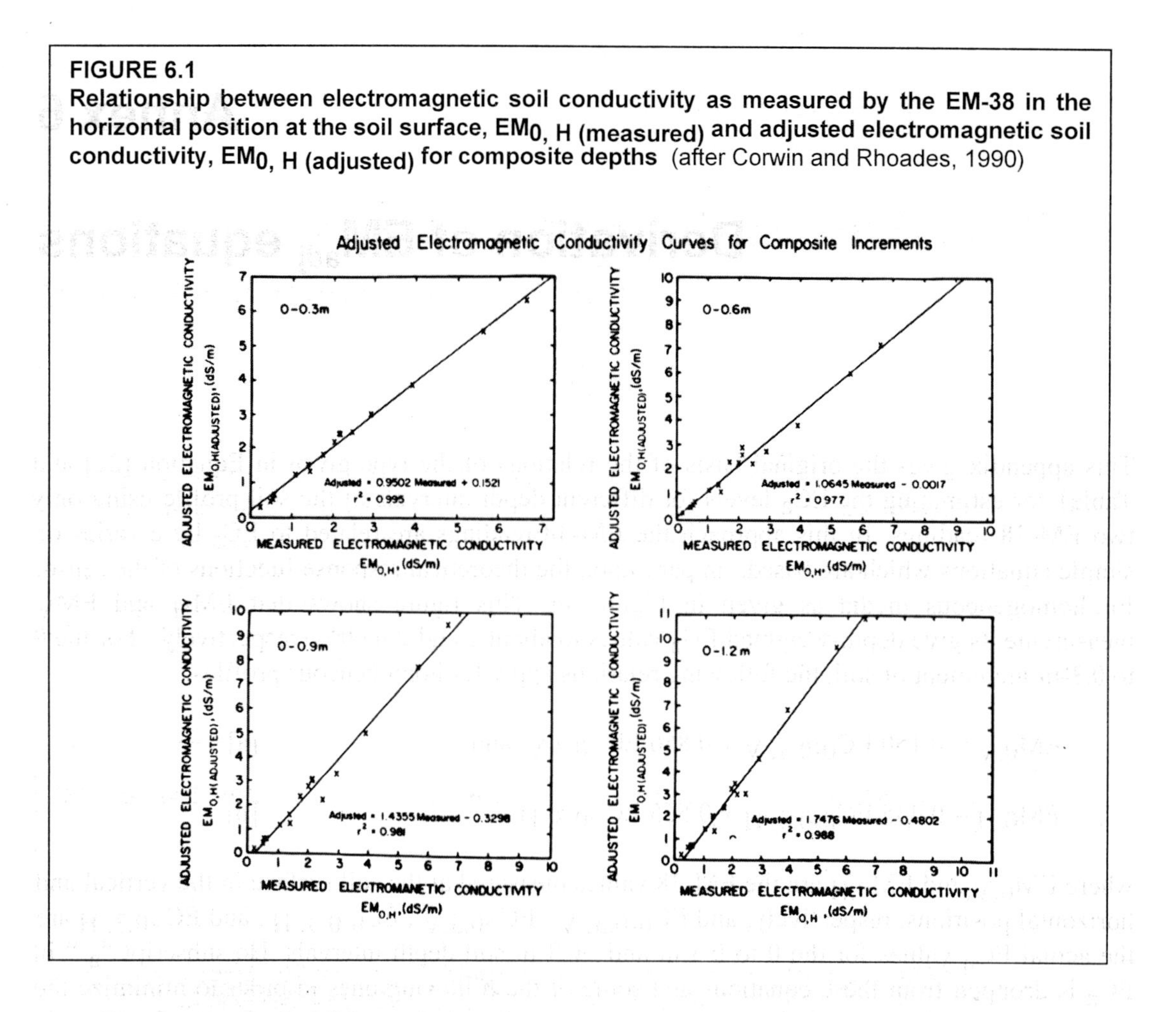

$$EM_{0,V} = 0.150\ EC_{0\text{-}0.3} + 0.850\ EC_{>0.3,V} \text{ and} \qquad [c]$$

$$EM_{0,H\ (adjusted,\ 0\text{-}0.3\text{-}m)} = 0.435\ EC_{0\text{-}0.3} + 0.565\ EC_{>0.3,\ V}. \qquad [d]$$

Equations [c] and [d] can now be reduced by substitution to form the following single equation:

$$EC_{a,\ 0\text{-}0.3} = 2.982\ EM_{0,H\ (adjusted,\ 0\text{-}0.3\text{-}m)} - 1.982\ EM_{0,V}\ , \qquad [e]$$

where $EM_{0,H\ (adjusted,\ 0\text{-}0.3\text{-}m)}$ is an empirically obtained linear expression of the form shown in Figure 94, such as:

$$EM_{0,H\ (adjusted,\ 0\text{-}0.3\text{-}m)} = k_1\ EM_{0,H} + k_2. \qquad [f]$$

Equations [e] and [f] can be reduced by substitution to form a single equation. Following the same rationale, an analogous set of equations can be obtained to predict EC_a for other soil depth intervals from $EM_{0,V}$ and $EM_{0,H}$. These equations are of the form:

$$EC_{a,\ x1\text{-}x2} = k_H\ EM_{0,H} - k_V\ EM_{0,V} + k\ , \qquad [g]$$

where k_H , k_V, and k are empirically determined coefficients for the depth interval x_1-x_2. The value of k should ideally be zero, but often is not, due to experimental error in the data. Equation [22] is a modification that was undertaken to achieve a more normal distribution of data so as to establish better values of the coefficients by statistical methods. Equation [23] is a further modification that was undertaken for the reasons explained in the main text.

Annex 7

Device for positioning EM-38 sensor during hand-held measurements

The device used to position the midpoints of the EM-38 magnets at a height of 10-cm and 50-cm above the ground when readings are taken in either the vertical or horizontal configurations is shown in Figure (7.1A and 7.1B) and Figure (7.2A and 7.2B), respectively. It is a block of lightweight wood (redwood) of dimensions 8.5 by 8.5 by 50 cm in which a 3.5 cm wide groove is cut 4.7 cm deep along the longitudinal axis and 6.0 cm deep in the end section (see Figure 7.3). A handle is attached by wooden dowels, plastic screws and glue; metal is not used or contained in any part of the "block". The EM-38 is laid on its side for the EM_H reading and is placed in the slot for the EM_V reading.

FIGURE 7.1
EM-38 sensor (centre of coils) positioned 10 cm above ground in the (A) vertical and (B) horizontal orientations, respectively

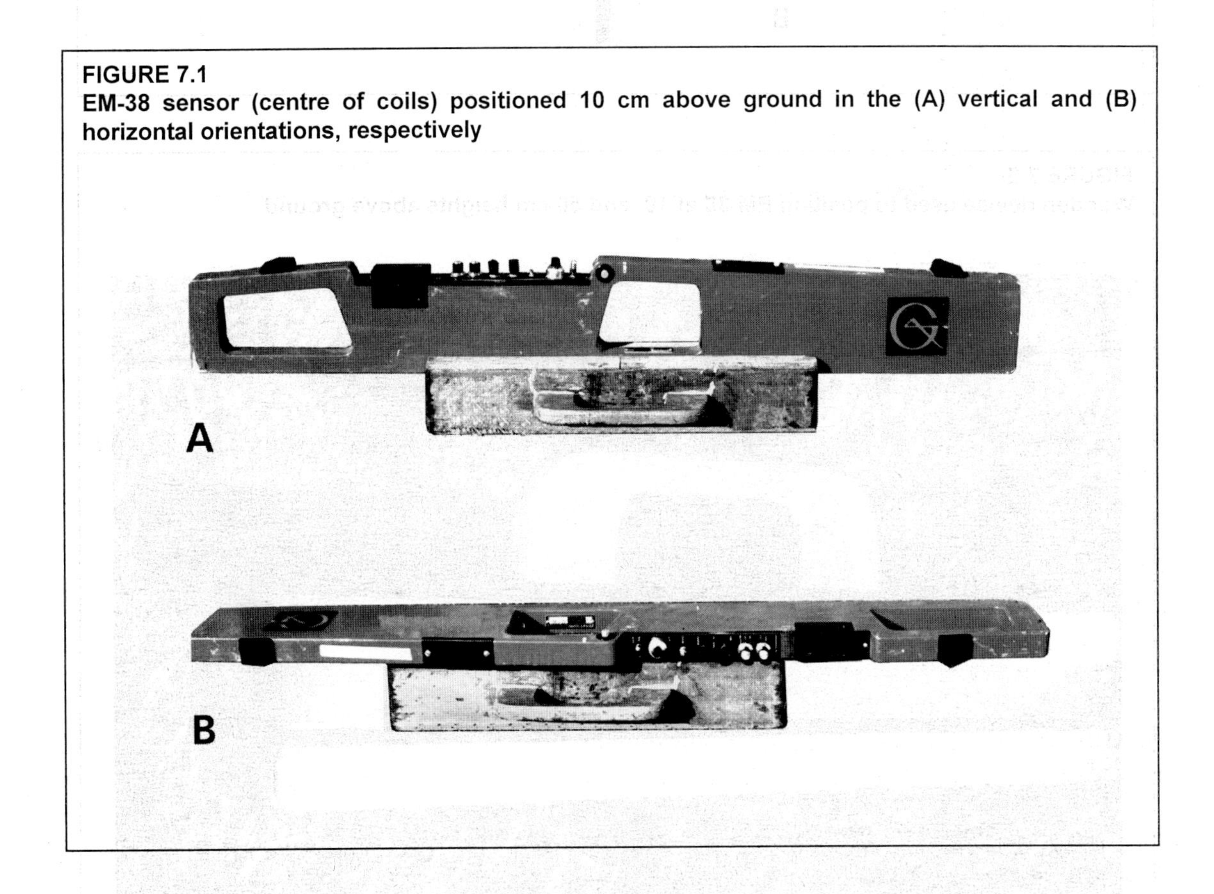

FIGURE 7.2
EM-38 sensor (centre of coils) positioned 50 cm above ground in the (A) vertical and (B) horizontal orientations, respectively

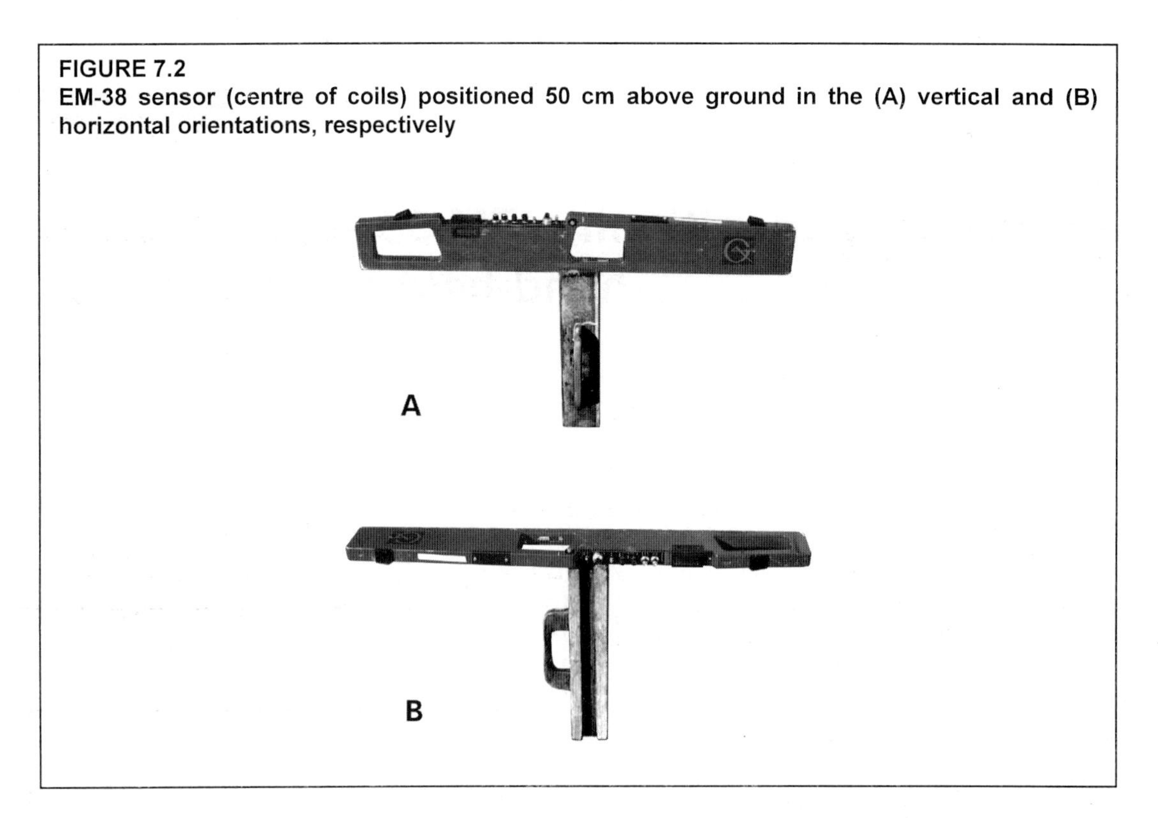

FIGURE 7.3
Wooden device used to position EM-38 at 10- and 50-cm heights above ground

Annex 8

Schematic and parts-list for soil four-electrode probe

The construction details of a mechanically fabricated soil EC-probe are provided in Figure 8-1. Four brass annular rings (electrodes) are juxtaposed between lucite insulators/spacers to form the probe. Teflon gaskets are placed between the electrodes and insulators/spacers, along with epoxy sealer, to prevent water from entering and shorting out the electrodes (see Figure 8.2A). The size of the probe is dimensioned to permit EC_a to be measured in 15-cm increments. The probe is afixed to a thick-walled, anodized aluminum shaft so that it can be inserted to the desired depth in the soil via a hole made with a standard 2.3 cm Oakfield or Lord soil sampler (see Figure 8.2B). The probe is slightly tapered (1°) toward the tip so that all four electrodes firmly contact the soil upon insertion in the hole. The leads from the electrodes exit the handle for connection to the generator/meter. More details and description of this soil EC-probe is given in Rhoades and van Schilfgaarde (1976). The commercial probe sold by the Eijkelkamp Agrisearch Equipment Company is essentially a copy of the above-described unit; that sold by Martek Instruments is a plastic-moulded version.

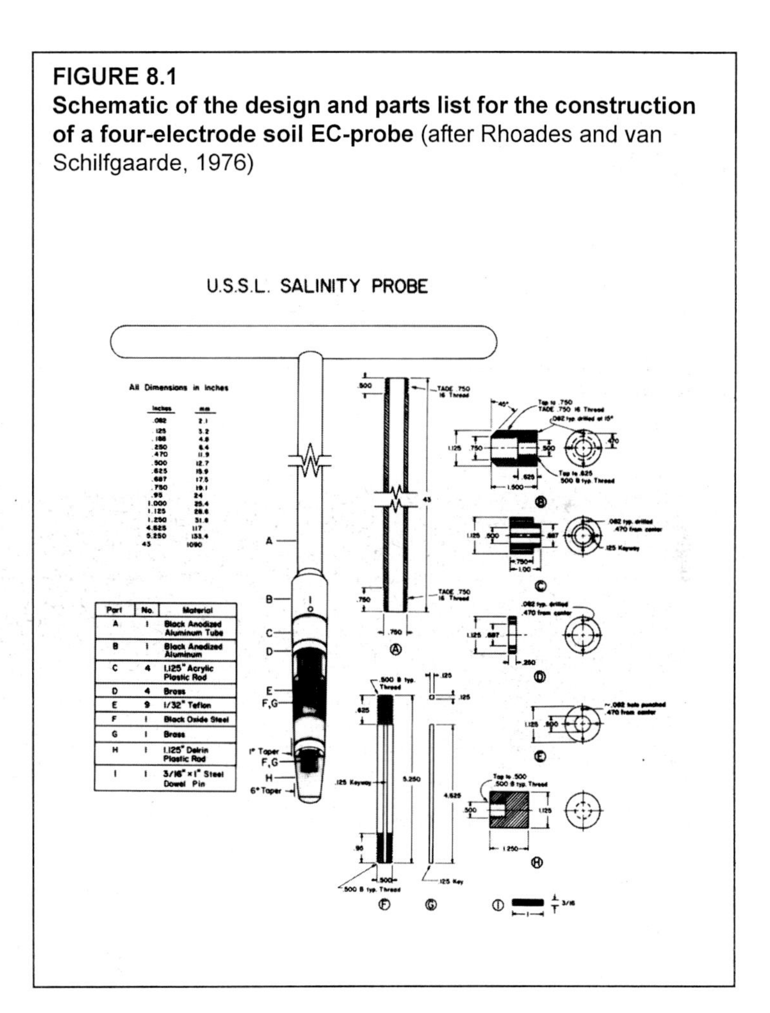

Inches	mm
.082	2.1
.125	3.2
.188	4.8
.250	6.4
.470	11.9
.500	12.7
.625	15.9
.687	17.5
.750	19.1
.95	24
1.000	25.4
1.125	28.6
1.250	31.8
4.625	117
5.250	133.4
43	1090

Part	No.	Material
A	1	Black Anodized Aluminum Tube
B	1	Black Anodized Aluminum
C	4	1.125" Acrylic Plastic Rod
D	4	Brass
E	9	1/32" Teflon
F	1	Black Oxide Steel
G	1	Brass
H	1	1.125" Delrin Plastic Rod
I	1	3/16" x 1" Steel Dowel Pin

FIGURE 8.1
Schematic of the design and parts list for the construction of a four-electrode soil EC-probe (after Rhoades and van Schilfgaarde, 1976)

FIGURE 8.2
(A) An unassembled and (B) assembled four-electrode soil EC-probe

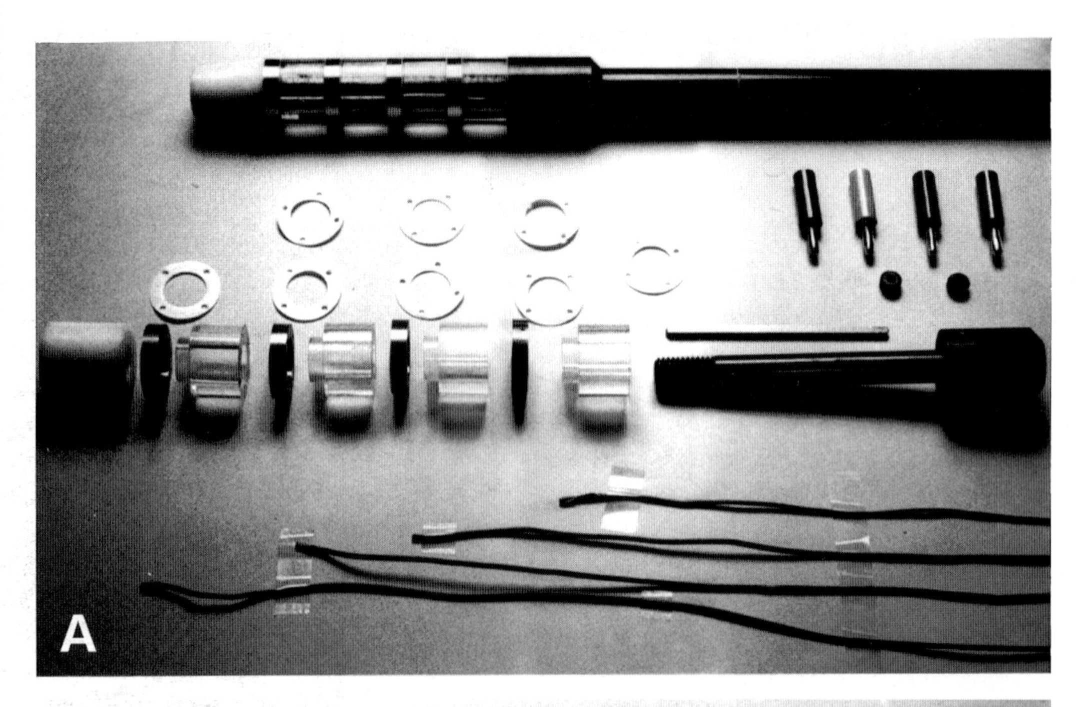

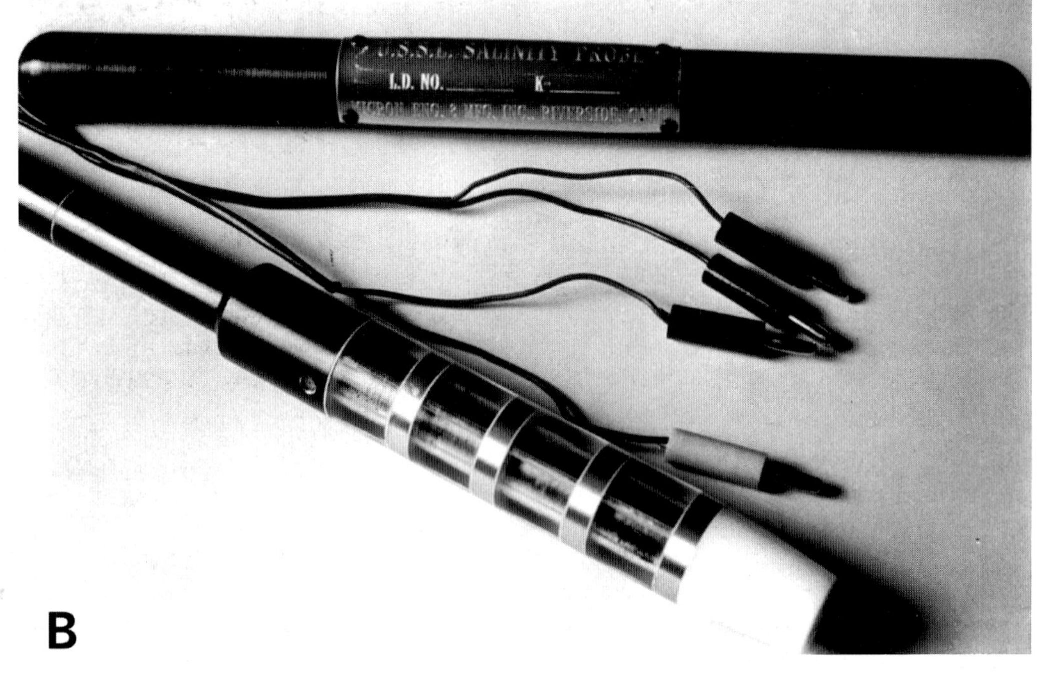

Annex 9

Description of statistical tests for monitoring soil salinity

The periodic assessment of soil salinity conditions over time is a critical component of any sort of serious, long-term monitoring strategy. The selection and acquisition of soil samples should be performed in a manner which optimizes the possibility of detecting any temporal and/or spatial-temporal trends occurring in the field. Lesch *et al.*, 1998, describes a regression based testing methodology that can be used to detect such trends. In this approach, and electromagnetic (EM) survey is performed at grid-points across the field and a limited number of soil salinity samples are then acquired at a selected small number of these survey grid sites. An EM-salinity regression model (Equation [34]) is then estimated and used to calculate (i.e., predict) the soil salinity levels at every grid site. Once this model has been estimated, new salinity samples can be acquired in the future at one or more of the known survey grid sites and compared to the predicted salinity levels (from the model). Lesch *et al.*, 1998, show that two statistical tests can be developed from such data: (1) a test to detect dynamic spatial variation over time - i.e., has the pattern changed in a non-random, dynamic manner across the field, and (2) a test to detect a shift in the median field salinity level - i.e., has the median salinity level increased up or decreased over time. This methodology does not require extensive soil sampling. For example, 12 to 20 soil samples are usually sufficient for establishing the initial regression equation, and the periodic acquisition of 8 to10 new soil samples are typically sufficient for testing purposes.

In more formal terms, the above tests are based on fitting a conditional regression model to the EM-salinity data, where the EM covariate data is assumed to be random (i.e., implying that the regression equation is really a mixed linear model). To do so, define y_{1j} and y_{2k} as the observed ln salinity levels from the jth and k^{th} sample site acquired during the first and second time frames, where $j = 1, 2, ..., n$ and $k = 1, 2, ..., m$. Let $\mathbf{y}_1$ represent the vector of observations from the first time frame, and $\mathbf{y}_2$ represent the observations from the second time frame. Additionally, define $\mathbf{X}_1$ as the matrix (grid) of EM covariate signal data observed during the first time frame. For this discussion, suppose that a survey grid of size N ($N > n,m$) of representative EM covariate data has been acquired during the first time frame only, and that the n and m sample sites (from the first and second time frames, respectively) are chosen from this grid. Note that the two sets of sample sites need not be collocated. Furthermore, assume that the conditional distribution of $\mathbf{y}$ given the observed $\mathbf{X}_1$ matrix is Normal with constant variance, and that the distribution of $\mathbf{X}_1$ does not depend on either the β parameter vector or the error variance. Assume that a suitable model for the first time frame is:

$$\mathbf{y}_1 \mid \mathbf{X}_1 = \mathbf{X}_1\beta + \epsilon_1 \qquad \text{(a)}$$

where $\in_1 \sim N(\mathbf{0},\sigma^2\mathbf{I})$, and $\mathbf{I}$ represents the identity matrix. Furthermore, assume that a suitable model for the second time frame is:

$$y_2 \mid \mathbf{X}_1 = \mathbf{X}_1\beta + \mathbf{d}_0 + \eta + \in_2 \qquad \text{(b)}$$

where $\mathbf{d}_0 = [d_0, d_0, \ldots, d_0]$, $\eta \sim N(\mathbf{0},\theta^2\mathbf{I})$, $\in_2 \sim N(0,\sigma^2\mathbf{I})$, and the η, $\in_1$, and $\in_2$ random error components are mutually independent. Hence, $(\mathbf{y}_2 \mid \mathbf{X}_1) - (\mathbf{y}_1 \mid \mathbf{X}_1) = \mathbf{d}_0 + \eta + \in_2 - \in_1$, where d_0 represents the shift in the average ln salinity level between the two time frames and the additional error term η represents the dynamic variability component.

Now, suppose that m new samples located on the grid are acquired during the *second time frame*. Let $\mathbf{y}_2$ represent this vector of sample observations, $\hat{y}_1$ represent the corresponding vector of predicted levels computed from equation (1) at these m sites, and define $\mathbf{H}_m = \mathbf{X}_m{}^T(\mathbf{X}_n{}^T\mathbf{X}_n)^{-1}\mathbf{X}_m$, where $\mathbf{X}_m$ represents the matrix of EM covariate data associated with these m prediction sites from the *first time frame*. Then equation (2) implies that the *prediction error* associated with these sites would be $(\mathbf{y}_2\text{-}\hat{\mathbf{y}}_1) \mid \mathbf{X}_1 \sim N(\mathbf{d}_0, \theta^2\,\mathbf{I}_m + \theta^2(\mathbf{I}_m + \mathbf{H}_m))$. Now, let $\mathbf{d} = \mathbf{y}_2\text{-}\hat{\mathbf{y}}_1$ where $= \{d_1, d_2, \ldots, d_m\}$. Define the calculated sample mean and variance of these observed differences as $\bar{u}$ and w^2, where $\bar{u} = (1/m)[d_1+d_2+\ldots+d_m]$ and $w^2 = (1/(m-1))[(d_1-\bar{u})^2+(d_2-\bar{u})^2+\ldots+(d_m-\bar{u})^2]$. Clearly $\bar{u}$ represents a conditionally unbiased estimate of d_0. Furthermore, given the previously stated modelling assumptions, the following three results can be derived: 1) an F-test for determining if $\theta^2 > 0$, 2) a method of moments estimate of θ^2, and 3) an approximate t-test for determining if $d_0 = 0$. These results are given below:

1. An F-test for determining if $\theta^2 > 0$ can be computed as $\phi = (\mathbf{d}\text{-}\bar{\mathbf{u}})^T\Sigma^{-1}(\mathbf{d}\text{-}\bar{\mathbf{u}}) \,/\, (m\text{-}1)s^2$, where $\Sigma = (\mathbf{I} + \mathbf{H}_m)$, and where ϕ is compared to an F distribution with m-1 and n-p-1 degrees of freedom.

2. The expected value of w^2 is $\theta^2 + \sigma^2(1+\lambda_1-\lambda_2)$, with $\lambda_1 = (1/m)\Sigma\, h_{ii}$ and $\lambda_2 = (1/(m(m-1)))\Sigma\Sigma\, h_{ij}\ \forall\ i \neq j$ (where h_{ij} represent the ith, j^{th} diagonal element of the $\mathbf{H}_m$ matrix). Hence, a method of moments estimate of θ^2 is $v^2 = w^2 - s^2(1+\lambda_1-\lambda_2)$.

3. An approximate t-test for $d_0 = 0$ can be computed as $c = \bar{u} \,/\, g$, where $g^2 = (1/m)v^2 + 2s^2[(1/m) + h_{mu}]$, $h_{mu} = \mathbf{x}_{mu}{}^T(\mathbf{X}_n{}^T\mathbf{X}_n)^{-1}\mathbf{x}_{mu}$, $\mathbf{x}_{mu} = (1/m)[\mathbf{x}_1 + \mathbf{x}_2 + \ldots + \mathbf{x}_m]$, and where c is compared to a t distribution with n-p-1 degrees of freedom. This test statistic assumes that the two sets of soil samples are not collocated.

Note that the F-test represents a test for dynamic salinity variation, while the t-test represents a test for a shift in the median level over time.

FAO TECHNICAL PAPERS

FAO IRRIGATION AND DRAINAGE PAPERS

1	Irrigation practice and water management, 1972 (Ar* E* F* S*)
1 Rev.1	Irrigation practice and water management, 1984 (E)
2	Irrigation canal lining, 1971 (New edition, 1977, available in E, F and S in the FAO Land and Water Development Series, No. 1)
3	Design criteria for basin irrigation systems, 1971 (E*)
4	Village irrigation programmes – a new approach in water economy, 1971 (E* F*)
5	Automated irrigation, 1971 (E* F* S*)
6	Drainage of heavy soils, 1971 (E* F* S*)
7	Salinity seminar, Baghdad, 1971 (E* F*)
8	Water and the environment, 1971 (E* F* S*)
9	Drainage materials, 1972 (E* F* S*)
10	Integrated farm water management, 1971 (E* F* S*)
11	Planning methodology seminar, Bucharest, 1972 (E* F*)
12	Farm water management seminar, Manila, 1972 (E*)
13	Water use seminar, Damascus, 1972 (E* F*)
14	Trickle irrigation, 1973 (E* F* S*)
15	Drainage machinery, 1973 (E* F*)
16	Drainage of salty soils, 1973 (C* E* F* S*)
17	Man's influence on the hydrological cycle, 1973 (E* F* S*)
18	Groundwater seminar, Granada, 1973 (E* F S*)
19	Mathematical models in hydrology, 1973 (E*)
20/1	Water laws in Moslem countries – Vol. 1, 1973 (E* F*)
20/2	Water laws in Moslem countries – Vol. 2, 1978 (E F)
21	Groundwater models, 1973 (E*)
22	Water for agriculture – index, 1973 (E/F/S*)
23	Simulation methods in water development, 1974 (E* F* S*)
24	Crop water requirements, (rev.) 1977 (C* E F S)
25	Effective rainfall, 1974 (C* E* F* S*)
26/1	Small hydraulic structures – Vol. 1, 1975 (E* F* S*)
26/2	Small hydraulic structures – Vol. 2, 1975 (E* F* S*)
27	Agro-meteorological field stations, 1976 (E* F* S*)
28	Drainage testing, 1976 (E* F* S*)
29	Water quality for agriculture, 1976 (E* F* S*)
29 Rev.1	Water quality for agriculture, 1985 (C** E* F* S*)
30	Self-help wells, 1977 (E*)
31	Groundwater pollution, 1979 (C* E* S)
32	Deterministic models in hydrology, 1979 (E*)
33	Yield response to water, 1979 (C* E F S)
34	Corrosion and encrustation in water wells, 1980 (E*)
35	Mechanized sprinkler irrigation, 1982 (C E* F S*)
36	Localized irrigation, 1980 (Ar C E* F S*)
37	Arid zone hydrology, 1981 (C E*)
38	Drainage design factors, 1980 (Ar C E F S)
39	Lysimeters, 1982 (C E* F* S*)
40	Organization, operation and maintenance of irrigation schemes, 1982 (C E* F S*)
41	Environmental management for vector control in rice fields, 1984 (E* F* S*)
42	Consultation on irrigation in Africa, 1987 (E F)
43	Water lifting devices, 1986 (E F)
44	Design and optimization of irrigation distribution networks, 1988 (E F)
45	Guidelines for designing and evaluating surface irrigation systems, 1989 (E*)
46	CROPWAT – a computer program for irrigation planning and management, 1992 (E F* S*)
47	Wastewater treatment and use in agriculture, 1992 (E*)
48	The use of saline waters for crop production, 1993 (E)
49	CLIMWAT for CROPWAT, 1993 (E)
50	Le pompage éolien, 1994 (F)
51	Prospects for the drainage of clay soils, 1995 (E)
52	Reforming water resources policy, 1995 (E)
53	Environmental impact assessment of irrigation and drainage projects, 1995 (E)
54	Crues et apports, 1996 (F)
55	Control of water pollution from agriculture, 1996 (E* S)
56	Crop evapotranspiration, 1998 (E)
57	Soil salinity assessment, 1999 (E)

Availability: May 1999

Ar –	Arabic	Multil –	Multilingual
C –	Chinese	*	Out of print
E –	English	**	In preparation
F –	French		
P –	Portuguese		
S –	Spanish		

The FAO Technical Papers are available through the authorized FAO Sales Agents or directly from Sales and Marketing Group, FAO, Viale delle Terme di Caracalla, 00100 Rome, Italy.

FAO TECHNICAL PAPERS

FAO IRRIGATION AND DRAINAGE PAPERS

1	Irrigation practice and water management, 1972 (Ar* E* F* S*)
1 Rev.1	Irrigation practice and water management, 1984 (E)
2	Irrigation canal lining, 1971 (New edition, 1977, available in E, F and S in the FAO Land and Water Development Series, No. 1)
3	Design criteria for basin irrigation systems, 1971 (E*)
4	Village irrigation programmes – a new approach in water economy, 1971 (E* F)
5	Automated irrigation, 1971 (E* F* S*)
6	Drainage of heavy soils, 1971 (E* F S*)
7	Salinity seminar, Baghdad, 1971 (E* F)
8	Water and the environment, 1971 (E* F* S*)
9	Drainage materials, 1972 (E* F* S*)
10	Integrated farm water management, 1971 (E* F* S*)
11	Planning methodology seminar, Bucharest, 1972 (E* F*)
12	Farm water management seminar, Manila, 1972 (E*)
13	Water use seminar, Damascus, 1972 (E* F*)
14	Trickle irrigation, 1973 (E* F* S*)
15	Drainage machinery, 1973 (E* F*)
16	Drainage of salty soils, 1973 (C* E* F* S*)
17	Man's influence on the hydrological cycle, 1973 (E* F* S*)
18	Groundwater seminar, Granada, 1973 (E* F S)
19	Mathematical models in hydrology, 1973 (E)
20/1	Water laws in Moslem countries – Vol. 1, 1973 (E* F)
20/2	Water laws in Moslem countries – Vol. 2, 1978 (E F)
21	Groundwater models, 1973 (E)
22	Water for agriculture – index, 1973 (E/F/S*)
23	Simulation methods in water development, 1974 (E F* S*)
24	Crop water requirements, (rev.) 1977 (C* E F S)
25	Effective rainfall, 1974 (C* E* F* S*)
26/1	Small hydraulic structures – Vol. 1, 1975 (E F S)
26/2	Small hydraulic structures – Vol. 2, 1975 (E F S)
27	Agro-meteorological field stations, 1976 (E F* S*)
28	Drainage testing, 1976 (E F S)
29	Water quality for agriculture, 1976 (E* F* S*)
29 Rev.1	Water quality for agriculture, 1985 (C** E F S)
30	Self-help wells, 1977 (E)
31	Groundwater pollution, 1979 (C* E S)
32	Deterministic models in hydrology, 1979 (E)
33	Yield response to water, 1979 (C* E F S)
34	Corrosion and encrustation in water wells, 1980 (E)
35	Mechanized sprinkler irrigation, 1982 (C E F S)
36	Localized irrigation, 1980 (Ar C E F S)
37	Arid zone hydrology, 1981 (C** E*)
38	Drainage design factors, 1980 (Ar C E F S)
39	Lysimeters, 1982 (C E F S)
40	Organization, operation and maintenance of irrigation schemes, 1982 (C** E F S**)
41	Environmental management for vector control in rice fields, 1984 (E F S)
42	Consultation on irrigation in Africa, 1987 (E F)
43	Water lifting devices, 1986 (E)
44	Design and optimization of irrigation distribution networks, 1988 (E F)
45	Guidelines for designing and evaluating surface irrigation systems, 1989 (E)
46	CROPWAT – a computer program for irrigation planning and management, 1992 (E F S)
47	Wastewater treatment and use in agriculture, 1992 (E)
48	The use of saline waters for crop production, 1993 (E)
49	CLIMWAT for CROPWAT, 1993 (E)
50	Le pompage éolien, 1994 (F)
51	Prospects for the drainage of clay soils, 1995 (E)
52	Reforming water resources policy – A guide to methods, processes and practices, 1995 (E)
53	Environmental impact assessment of irrigation and drainage projects, 1995 (A)
54	Crues et apports, 1996 (F)
55	Control of water pollution from agriculture, 1996 (E S)
56	Crop evapotranspiration – Guidelines for computing crop water requirements, 1998 (E)
57	Soil salinity assessment – Methods and interpretation of electrical conductivity measurements, 1999 (E)

Availability: May 1999

Ar –	Arabic	Multil –	Multilingual
C –	Chinese	*	Out of print
E –	English	**	In preparation
F –	French		
P –	Portuguese		
S –	Spanish		

The FAO Technical Papers are available through the authorized FAO Sales Agents or directly from Sales and Marketing Group, FAO, Viale delle Terme di Caracalla, 00100 Rome, Italy.